Smart CMOS Image Sensors and Applications

OPTICAL SCIENCE AND ENGINEERING
Founding Editor **Brian J. Thompson**
University of Rochester Rochester, New York

Nonlinear Optics: Theory, Numerical Modeling, and Applications, *Partha P. Banerjee*

Semiconductor Laser Fundamentals, *Toshiaki Suhara*

High-Performance Backbone Network Technology, *edited by Naoaki Yamanaka*

Organic Light-Emitting Diodes: Principles, Characteristics, and Processes, *Jan Kalinowski*

Micro-Optomechatronics, *Hiroshi Hosaka, Yoshitada Katagiri, Terunao Hirota, and Kiyoshi Itao*

Microoptics Technology: Second Edition, *Nicholas F. Borrelli*

Organic Photovoltaics: Mechanism, Materials, and Devices, *edited by Sam-Shajing Sun and Niyazi Serdar Saracftci*

Physics of Optoelectronics, *Michael A. Parker*

Image Sensors and Signal Processor for Digital Still Cameras, *Junichi Nakamura*

GMPLS Technologies: Broadband Backbone Networks and Systems, *Naoaki Yamanaka, Kohei Shiomoto, and Eiji Oki*

Electromagnetic Theory and Applications for Photonic Crystals, *Kiyotoshi Yasumoto*

Encyclopedic Handbook of Integrated Circuits, *edited by Kenichi Iga and Yasuo Kokubun*

Laser Safety Management, *Ken Barat*

Optics in Magnetic Multilayers and Nanostructures, *Štefan Višňovský*

Applied Microphotonics, *edited by Wes R. Jamroz, Roman Kruzelecky, and Emile I. Haddad*

Polymer Fiber Optics: Materials, Physics, and Applications, *Mark G. Kuzyk*

Lens Design, Fourth Edition, *Milton Laikin*

Photonics: Principles and Practices, *Abdul Al-Azzawi*

Gas Lasers, *edited by Masamori Endo and Robert F. Walker*

Organic Field-Effect Transistors, *Zhenan Bao and Jason Locklin*

Coarse Wavelength Division Multiplexing: Technologies and Applications, *edited by Hans Joerg Thiele and Marcus Nebeling*

Terahertz Spectroscopy: Principles and Applications, *edited by Susan L. Dexheimer*

Slow Light: Science and Applications, *edited by Jacob B. Khurgin and Rodney S. Tucker*

Near-Earth Laser Communications, *edited by Hamid Hemmati*

Photoacoustic Imaging and Spectroscopy, *edited by Lihong V. Wang*

Handbook of Optical and Laser Scanning, Second edition, *Gerald F. Marshall, Glenn E. Stutz*

Tunable Laser Applications, Second Edition, *edited by F. J. Duarte*

Laser Beam Shaping Applications, Second edition, *Fred M. Dickey, Todd E. Lizotte*

Lightwave Engineering, *Yasuo Kokubun*

Laser Safety: Tools and Training, *edited by Ken Barat*

Computational Methods for Electromagnetic and Optical Systems, *John M. Jarem, Partha P. Banerjee*

Biochemical Applications of Nonlinear Optical Spectroscopy, *edited by Vladislav Yakovlev*

Optical Methods of Measurement: Wholefield Techniques, Second Edition, *Rajpal Sirohi*

Fundamentals and Basic Optical Instruments, *Daniel Malacara Hernández*

Advanced Optical Instruments and Techniques, *Daniel Malacara Hernández*

Entropy and Information Optics: Connecting Information and Time, Second Edition, *Francis T.S. Yu*

Handbook of Optical Engineering, Second Edition, Two Volume Set, *Daniel Malacara Hernández*

Optical Materials and Applications, *Moriaki Wakaki*

Photonic Signal Processing: Techniques and Applications, *Le Nguyen Binh*

Practical Applications of Microresonators in Optics and Photonics, *edited by Andrey B. Matsko*

Near-Earth Laser Communications, Second Edition, *edited by Hamid Hemmati*

Nonimaging Optics: Solar and Illumination System Methods, Design, and Performance, *Roland Winston, Lun Jiang, Vladimir Oliker*

Smart CMOS Image Sensors and Applications, *Jun Ohta*

For more information about this series, please visit: https://www.crcpress.com/Optical-Science-and-Engineering/book-series/CRCOPTSCIENG

Smart CMOS Image Sensors and Applications

Jun Ohta

CRC Press
Taylor & Francis Group
Boca Raton   London   New York

CRC Press is an imprint of the
Taylor & Francis Group, an **informa** business

Second edition published 2020
by CRC Press
6000 Broken Sound Parkway NW, Suite 300, Boca Raton, FL 33487-2742

and by CRC Press
4 Park Square, Milton Park, Abingdon, Oxon OX14 4RN

First issued in paperback 2023

First edition published by CRC Press 2008

© 2002 by Taylor & Francis Group, LLC
CRC Press is an imprint of Taylor & Francis Group, an Informa business

No claim to original U.S. Government works

ISBN 13: 978-1-03-265236-8 (pbk)
ISBN 13: 978-1-4987-6464-3 (hbk)
ISBN 13: 978-1-315-15625-5 (ebk)

DOI: 10.1201/9781315156255

Visit the Taylor & Francis Web site at
http://www.taylorandfrancis.com

and the CRC Press Web site at
http://www.crcpress.com

Contents

Preface to the Second Edition   **ix**

About the Author   **xiii**

1 Introduction   **1**
   1.1   A general overview 1
   1.2   Brief history of CMOS image sensors 2
      1.2.1   Competition with CCDs 3
      1.2.2   Solid-state imagers with in-pixel amplification 4
      1.2.3   Present CMOS image sensors 4
   1.3   Brief history of smart CMOS image sensors 5
      1.3.1   Vision chips 5
      1.3.2   Advancement of CMOS technology and smart CMOS
           image sensors 7
      1.3.3   Smart CMOS image sensors based on high performance
           CMOS image sensor technologies 7
   1.4   Organization of the book 9

2 Fundamentals of CMOS image sensors   **11**
   2.1   Introduction 11
   2.2   Fundamentals of photo-detection 12
      2.2.1   Absorption coefficient 12
      2.2.2   Behavior of minority carriers 13
      2.2.3   Sensitivity and quantum efficiency 15
   2.3   Photo-detectors for smart CMOS image sensors 16
      2.3.1   pn-junction photodiode 16
      2.3.2   Photo-gate 25
      2.3.3   Photo-transistor 26
      2.3.4   Avalanche photodiode 26
      2.3.5   Photo-conductive detector 28
   2.4   Accumulation mode in PDs 29
      2.4.1   Potential change in accumulation mode 30
      2.4.2   Potential description 31
      2.4.3   Behavior of photo-generated carriers in PD 32
   2.5   Basic pixel structures 36
      2.5.1   Passive pixel sensor 36

	2.5.2	Active pixel sensor, 3T-APS	38
	2.5.3	Active pixel sensor, 4T-APS	40
2.6	Sensor peripherals	43
	2.6.1	Addressing	43
	2.6.2	Readout circuits	45
	2.6.3	Analog-to-digital converters	48
2.7	Basic sensor characteristics	50
	2.7.1	Noise	50
	2.7.2	Dynamic range	52
	2.7.3	Speed	53
2.8	Color	54
	2.8.1	On-chip color filter type	54
	2.8.2	Three imagers type	56
	2.8.3	Three light sources type	56
2.9	Comparison among pixel architectures	57
2.10	Comparison with CCDs	57

3 Smart structures and materials   61

3.1	Introduction	61
3.2	Smart pixels	62
	3.2.1	Analog mode	62
	3.2.2	Pulse modulation mode	67
	3.2.3	Digital mode	76
3.3	Smart materials and structures	78
	3.3.1	Silicon-on-insulator	79
	3.3.2	Extending to NIR region	81
	3.3.3	Backside illumination	84
	3.3.4	3D integration	85
	3.3.5	Smart structure for color detection	87
3.4	Dedicated pixel arrangement and optics for smart CMOS image	
	sensors	92
	3.4.1	Phase-difference detection auto focus	93
	3.4.2	Hyper omni vision	93
	3.4.3	Biologically inspired imagers	94
	3.4.4	Light field camera	96
	3.4.5	Polarimetric imaging	99
	3.4.6	Lensless imaging	103

4 Smart imaging   107

4.1	Introduction	107
4.2	High sensitivity	108
	4.2.1	Dark current reduction	108
	4.2.2	Differential APS	110
	4.2.3	High conversion gain pixel	111
	4.2.4	SPAD	111

	4.2.5	Column-parallel processing	111
4.3	High-speed		114
	4.3.1	Overview	114
	4.3.2	Global shutter	115
	4.3.3	Column- and pixel-parallel processing for high speed imaging	117
	4.3.4	Ultra-high-speed	118
4.4	Wide dynamic range		122
	4.4.1	Overview	122
	4.4.2	Nonlinear response	125
	4.4.3	Linear response	126
4.5	Demodulation		130
	4.5.1	Overview	130
	4.5.2	Correlation	131
	4.5.3	Method of two accumulation regions	133
4.6	Three-dimensional range finder		137
	4.6.1	Overview	137
	4.6.2	Time-of-flight	139
	4.6.3	Triangulation	142

5 Applications **145**
5.1	Introduction		145
5.2	Information and communication technology applications		145
	5.2.1	Optical wireless communication	146
	5.2.2	Optical ID tag	155
5.3	Chemical applications		160
	5.3.1	Optical activity imaging	160
	5.3.2	pH imaging sensor	163
5.4	Bioscience and Biotechnology applications		165
	5.4.1	Attachment type	167
	5.4.2	On-chip type	169
	5.4.3	Implantation type	181
5.5	Medical applications		190
	5.5.1	Capsule endoscope	190
	5.5.2	Retinal prosthesis	192

Appendices **205**

A Tables of constants **207**

B Illuminance **209**

C Human eye and CMOS image sensors **213**

D Wavelength region in visible and infrared lights **217**

E   Fundamental characteristics of MOS capacitors   **219**

F   Fundamental characteristics of MOSFET   **221**

G   Optical format and resolution   **225**

H   Intrinsic optical signal and *in vivo* window   **227**

References   **229**

Index   **285**

Preface to the Second Edition

This book is the 2nd edition of *Smart CMOS Image Sensors and Applications* published in 2007. The purpose of coming out with this edition is essentially the same as that of the 1st edition. Here, I reiterate the aim of this book, which was first stated in the preface of the 1st edition.

> *Smart CMOS Image Sensors and Applications focuses on smart functions implemented in CMOS image sensors and their applications. Some sensors have already been commercialized, whereas some have only been proposed; the field of smart CMOS image sensors is active and generating new types of sensors. In this book I have endeavored to gather references related to smart CMOS image sensors and their applications; however, the field is so vast that it is likely that some topics are not described. Furthermore, the progress in the field is so rapid that some topics will develop as the book is being written. However, I believe the essentials of smart CMOS image sensors are sufficiently covered and that this book is therefore useful for graduate school students and engineers entering the field.*

It has been 12 years since the 1st edition of this book was published and during these long years, the environment surrounding image sensors has changed significantly. The smartphones, nowadays, are equipped with at least two cameras, one on the front side and another on the rear side. The performance parameters of smart phone cameras, such as the number of pixels, sensitivity, and speed, have improved dramatically. Some of the smart functions introduced in the 1st edition of this book have already been employed in commercial products. Based on these changes, the first edition has been revised to adapt the latest developments in smart CMOS image sensors and their applications to the current technology. However, there have been so many technological advances in this field over the past 12 years that it became more expedient to entirely rewrite some of the sections of the 1st edition.

The organization of this book is almost the same as that of the 1st edition. Firstly, Chapter 1 introduces MOS imagers and smart CMOS image sensors. Chapter 2 describes the basic elements of CMOS image sensors and details the relevant physics behind these optoelectronic devices. Typical CMOS image sensor structures, such as active pixel sensors (APS), are also introduced in this chapter. The subsequent chapters (i.e., 3 , 4 and 5) form the crux of the book, and delve into smart CMOS image sensors. Chapter 3 introduces several smart structures and materials for smart CMOS image sensors. Using these structures and materials, Chapter 4 describes smart imaging features, such as high sensitivity, high speed imaging, wide dynamic range image sensing, and three-dimensional (3D) range finding. In Chapter 5, which is the final chapter in this book, applications of smart CMOS image sensors are demonstrated in the fields of information and communication technology (ICT), chemistry, biology, and medicine

This work is inspired by numerous preceding books related to CMOS image sensors. In particular, A. Moini's, *Vision Chips* [1], which features a comprehensive archive of vision chips, J. Nakamura's, *Image Sensors and Signal Processing for Digital Still Cameras* [2], which presents recent rich results of this field, K. Yonemoto's introductory but comprehensive book on CCD and CMOS imagers, *Fundamentals and Applications of CCD/CMOS Image Sensors* [3], and T. Kuroda's book, *Essential Principles of Image Sensors* [4]. Of these, I was particularly impressed by K. Yonemoto's book, which unfortunately has only been published in Japanese. I hope that the present work helps to illuminate this field and that it complements that of Yonemoto's book. I have also been influenced by books written by numerous other senior Japanese researchers in this field, including Y. Takemura [5], Y. Kiuchi [6], T. Ando and H. Komobuchi [7], and K. Aizawa and T. Hamamoto [8]. The book on CCDs by A.J.P. Theuwissen is also useful [9].

I would like to thank all the people who have contributed both directly and indirectly to the areas covered in this book. I am particularly grateful to the current as well as past faculty members of my laboratory, the Laboratory of Photonic Device Science in the Division of Materials Science at the Nara Institute of Science and Technology (NAIST) for their meaningful and significant contributions, which predominantly constitute the contents of Chapter 5. The current faculty members include Prof. Kiyotaka Sasagawa, Prof. Hiroyuki Tashiro, Prof. Makito Haruta, and Prof. Hironari Takehara, while the former faculty members include Prof. Takashi Tokuda, currently of Tokyo Institute of Technology, Prof. Keiichiro Kagawa, currently of Shizuoka University, Prof. Toshihiko Noda, currently of Toyohashi University of Technology, and Prof. Hiroaki Takehara, currently of University of Tokyo. This book would not have seen the light of day without their efforts. I would also like to thank Prof. Emeritus Masahiro Nunoshita for his continuous encouragement in the early stages of the development of our laboratory, and Ms. Ryoko Fukuzawa and Ms. Kazumi Matsumoto, who, respectively, are the current and past secretaries of my laboratory, for their constant support in numerous administrative affairs. In addition, I would like to express my gratitude to the postdoctoral fellows, technical staff and the graduate students in my laboratory, both past and present, for their assistance and research efforts.

For topics related to retinal prosthesis, I would like to thank the late Prof. Yasuo Tano, who was the first project leader of the retinal prosthesis project, and Prof. Takashi Fujikado and Prof. Tetsuya Yagi at Osaka University. I would also like to thank the members of the retinal prosthesis project at Nidek Co. Ltd., particularly, Mr. Motoki Ozawa, Mr. Kenzo Shodo, Dr. Yasuo Terasawa, Dr. Hiroyuki Kanda, and Ms. Naoko Tsunematsu. In addition, I would like to thank the collaborators of the *in vivo* image sensor project, Prof. Emeritus Sadao Shiosaka, Prof. Hideo Tamura, Prof. David C. Ng, and Dr. Takuma Kobayashi. I would like to express my gratitude to Prof. Emeritus Masamitsu Haruna for his continuous encouragement for the research of the *in vivo* image sensors, and Prof. Emeritus Ryoichi Ito, the supervisor of my bachelor's, master's and doctoral theses, for his continuous encouragement throughout my research.

I first entered the research area of smart CMOS image sensors as a visiting researcher at the University of Colorado at Boulder under Prof. Kristina M. Johnson in 1992 to 1993. My experience there was very exciting and it helped me enormously in my initial research into smart CMOS image sensors after returning to Mitsubishi Electric Corporation. I would like to thank all of my colleagues at Mitsubishi Electric Corp. for their help and support, including Prof. Hiforumi Kimata, Dr. Shuichi Tai, Dr. Kazumasa Mitsunaga, Prof. Yutaka Arima, Prof. Masahiro Takahashi, Dr. Yoshikazu Nitta, Dr. Eiichi Funatsu, Dr. Kazunari Miyake, Mr. Takashi Toyoda, and numerous others.

I am most grateful to Prof. Chung-Yu Wu, Prof. Masatoshi Ishikawa, Prof. Emeritus Mitumasa Koyanagi, Prof. Jun Tanida, Prof. Shoji Kawahito, Prof. Atsushi Ono and Dr. C. Tubert, for courteously allowing me to use their figures and data in this book.

I have learned extensively from the committee members of the Institute of Image Information and Television Engineers, Japan, especially, Prof. Shigetoshi Sugawa, Prof. Kiyoharu Aizawa, Prof. Kazuaki Sawada, Prof. Takayuki Hamamoto, Prof. Junichi Akita, Prof. Rihito Kuroda, and the researchers of numerous image sensor groups in Japan, including Dr. Yasuo Takemura, Dr. Takao Kuroda, Prof. Nobukazu Teranishi, Dr. Hirofumi Sumi, and Dr. Junichi Nakamura. I would particularly like to thank Taisuke Soda, who initially presented me with the opportunity to write this book, and Mark Gutierrez of CRC Press, Taylor & Francis Group for his patience during the compilation of this book. Without their continuous encouragement, the publication of this book would not have been possible.

Personally, I extend my deepest thanks to the late Mr. Ichiro Murakami for continually stimulating my enthusiasm in image sensors and related topics, and the late Prof. Yasuo Tano for his support during the initial stage of my research at NAIST. I could not have carried out my research today without their supports. I would like to pray for their souls.

Finally, I would like to extend my special thanks to my wife, Dr. Yasumi Ohta, who also happens to be a postdoctoral fellow in my laboratory, for her support and understanding during the course of writing this book. Without her efforts, the research projects of *in vivo* imaging would not be successful.

Jun Ohta
Nara, Japan, March 2020

About the Author

Jun Ohta was born in Gifu, Japan in 1958. He received his B.E., M.E., and Dr. Eng. degrees in applied physics, all from the University of Tokyo, Japan, in 1981, 1983, and 1992, respectively. In 1983, he joined Mitsubishi Electric Corporation, Hyogo, Japan, where he has been engaged in research on optoelectronic integrated circuits, optical neural networks, and artificial retina chips. From 1992 to 1993, he was a visiting scientist in Optoelectronic Computing Systems Center, University of Colorado in Boulder. In 1998, he became an associate professor at the Graduate School of Materials Science, Nara Institute of Science and Technology (NAIST), Nara, Japan, where he subsequently, became a professor in 2004. His current research interests include smart CMOS image sensors, retinal prosthesis devices, and microelectronics-based biomedical devices and systems. Dr. Ohta was the recipient of a number of awards, which include the best paper award from IEICE Japan in 1992, Ichimura award in 1996, National Commendation for Invention in 2001, Niwa Takayanagi Award in 2007, Izuo Hayashi Award in 2009, IEICE Electronics Society Achievement Award in 2018, etc. He serves as the Distinguished Lecturer of IEEE Solid-State Circuits Society, a section editor of *IET Journal of Engineering*, an associate editor of *IEEE Trans. Biomedical Circuits and Systems*, an editor of *Japanese Journal of Applied Physics*, etc. He also served as technical program committee member of IEEE ISSCC, Symposium on VLSI Circuits, IEEE BioCAS 2018, etc. He organized the IEEE BioCAS 2019 at Nara as one of the general co-chairs. He is a senior member of both IEEE and the Institute of Electrical Engineers of Japan, and a fellow member of both the Japan Society of Applied Physics and the Institute of Image Information and Television Engineers of Japan.

1

Introduction

1.1 A general overview

Complementary metal-oxide-semiconductor (CMOS) image sensors have been the subject of extensive research and development and currently surpass the market with charge coupled device (CCD) image sensors, which have dominated the field of imaging sensors for a long time. CMOS image sensors are now widely used not only for consumer electronics, such as compact digital still cameras (DSC), mobile phone cameras, handy-camcorders, and digital single lens reflex (DSLR) cameras, but also for cameras used in automobiles, surveillance, security, robot vision, etc. Recently, further applications of CMOS image sensors in biotechnology and medicine have emerged. Many of these applications require advanced performance features, such as wide dynamic range, high speed, and high sensitivity, while others need dedicated functions, for example, three-dimensional (3D) range finding. It is difficult to perform such tasks with conventional image sensors. Furthermore, some signal processing devices are insufficient for these purposes. Smart CMOS image sensors or CMOS image sensors with integrated smart functions on the chip may meet the requirements of these applications.

CMOS image sensors are fabricated based on standard CMOS large scale integration (LSI) fabrication processes, while CCD image sensors are based on a specially developed fabrication process. This feature of CMOS image sensors makes it possible to integrate the functional circuits to develop smart CMOS image sensors and that can realize not only a higher performance than that of the CCD and conventional CMOS image sensors, but also perform versatile functions that cannot be achieved with conventional image sensors.

Smart CMOS image sensors are mainly aimed at two objectives: (i) enhancing or improving the fundamental characteristics of CMOS image sensors, such as dynamic range, speed, and sensitivity; (ii) implementing new functions, such as 3D range finding and modulated light detection. For both objectives, many architectures and/or structures, as well as materials, have been proposed and demonstrated.

The following terms are also associated with smart CMOS image sensors: computational CMOS image sensors, integrated functional CMOS image sensors, vision chips, focal plane image processing, as well as many others. With the exception of vision chips, these terms suggest that an image sensor has other functions in addition to imaging. The name vision chip originates from a device proposed and developed by C. Mead and coworkers, which mimics the human visual processing system. This topic will be described later in this chapter. In the following section, we first present a survey of the history of CMOS image sensors in general, and then briefly review the history of smart CMOS image sensors, in particular. For a summarized history of various imagers, including the CCDs, the readers are advised to refer to Ref. [10].

1.2   Brief history of CMOS image sensors

Birth of MOS imagers

The history of MOS image sensors, shown in Fig. 1.1, started with solid-sate imagers used as a replacement for image tubes. For solid-state image sensors, four important functions had to be realized: light-detection, accumulation of photo-generated signals, switching from accumulation to readout, and scanning. These functions are discussed in Chapter 2. The scanning function in X–Y addressed silicon-junction photosensing devices was proposed in the early 1960s by S.R. Morrison at Honeywell as the "photoscanner" [11] and by J.W. Horton *et al.* at IBM as the "scanistor" [12]. P.K. Weimer *et al.* proposed solid-state image sensors with scanning circuits using thin-film transistors (TFTs) [13]. In these devices, photo-conductive film, discussed in Sec. 2.3.5, is used for the photo-detector. M.A. Schuster and G. Strull at NASA used photo-transistors (PTrs) as photo-detectors, as well as switching devices to realize X–Y addressing [14]. They successfully obtained images with a fabricated 50 × 50-pixel array sensor. PTrs are discussed in Sec. 2.3.3.

The accumulation mode in a photodiode is an important function for MOS image sensors and is described in Sec. 2.4. It was first proposed by G.P. Weckler at Fairchild Semiconductor [15]. In the proposal, the floating source of a metal–oxide–semiconductor filed–effect transistor (MOSFET) was used as a photodiode. This structure is used in some present CMOS image sensors. Weckler later fabricated and demonstrated a 100 × 100-pixel image sensor using this structure [16]. Since then, several types of solid-sate image sensors have been proposed and developed [16–19], as summarized in Ref. [20].

The solid-state image sensor developed by P.J. Noble at Plessey was almost the same as the MOS image sensor or passive pixel sensor (PPS), discussed in Sec. 2.5.1, consisting of a photodiode and a switching MOS transistor in a pixel with X- and Y-scanners and a charge amplifier. Noble briefly discussed the possibility of integrating logic circuitry for pattern recognition on a chip, which may be the first prediction of a smart CMOS image sensor.

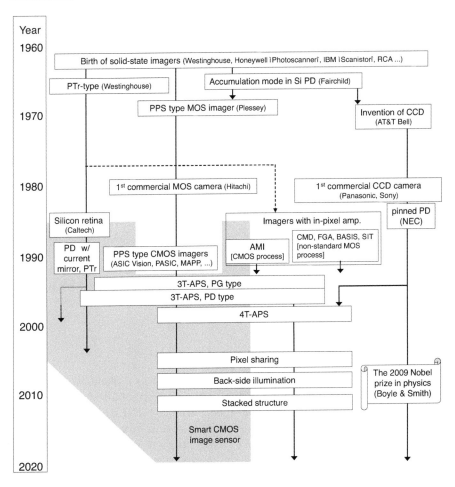

FIGURE 1.1: Evolution of MOS image sensors and related inventions.

1.2.1 Competition with CCDs

Shortly after the publication of the details of solid-state image sensors in *IEEE Transactions on Electron Devices* in 1968, the CCD image sensors become prominent [21]. The CCD itself was invented in 1969 by W. Boyle and G.E. Smith at AT&T Bell Laboratories [21] and was experimentally verified at almost the same time [22]. Initially, the CDD was developed as semiconductor memory, and as a replacement for magnetic bubble memory; however, it was soon developed for use in image sensors, which was invented by Michael F. Tompsett [23]. The early stage of the invention of the CCD is described in Ref. [24].

Considerable research effort resulted in the production of the first commercial MOS imagers, appearing in the 1980s [25–30]. While Hitachi had developed MOS imagers [25, 27], until recently only the CCDs were widely manufactured and used as they offered superior image quality than MOS imagers.

1.2.2   Solid-state imagers with in-pixel amplification

Subsequently, efforts were made to improve the signal-to-noise ratio (SNR) of MOS imagers by incorporating an amplification mechanism in a pixel. In the 1960s, a photo-transistor (PTr) type imager was developed [14]. In the late 1980s, several amplifier type imagers were developed, including the charge modulated device (CMD) [31], floating gate array (FGA) [32], base-stored image sensor (BASIS) [33], static induction transistor (SIT) [34], amplified MOS imager (AMI) [35–39], and some others [6, 7]. Apart from the AMI, these required some modification of standard MOS fabrication technology in the pixel structure. Ultimately they were not commercialized and their development was terminated. The AMI can be fabricated using the standard CMOS technology without any modification; however, its pixel structure is the same as that of the active pixel sensor (APS). It should be noted that the AMI uses an I–V converter as a readout circuit while the APS uses a source follower, even though this difference is not critical. The APS is also classified as an image sensor with in-pixel amplification.

1.2.3   Present CMOS image sensors

The APS was first realized by using a photogate (PG) as a photo-detector by E. Fossum *et al.* at JPL* and then by using a photodiode (PD) [40, 41]. A PG was used owing mainly to the ease of signal charge handling. The sensitivity of a PG is not adequate as poly-silicon as a gate material is opaque in the visible wavelength region. APSs using a PD are called 3T-APSs (three transistor APSs) and are now widely used in CMOS image sensors. In the first stage of 3T-APS development, the image quality could not compete with that of CCDs, both with respect to fixed pattern noise (FPN) and random noise. Introducing noise canceling circuits reduces the FPN but not the random noise.

By incorporating a pinned PD structure used in the CCDs, which has a low dark current and complete depletion structure, the 4T-APS (four transistor APS) was successfully developed [42]. A 4T-APS can be used with correlated double sampling (CDS), which can eliminate $k_B T C$ noise, which is the main factor in random noise. The image quality of a 4T-APS can compete with that of the CCDs. A significant concern with 4T-APSs is their large pixel size compared to that in the CCDs. The 4T-APS has four transistors plus a PD and floating diffusion (FD) in a pixel, while the CCD has one transfer gate plus a PD. Although the CMOS technology advances have benefited the development of CMOS image sensors [43], namely in shrinking the pixel size, it is essentially difficult to realize a smaller pixel size than that of the CCDs. Recently, a pixel sharing technique has been widely used in 4T-APSs that is

---

*Jet Propulsion Laboratory.

effective in reducing the pixel size to a value that is comparable to that of the CCDs. Figure 1.2 shows the trend of the pixel pitch in 4T-APSs. This figure illustrates that the pixel pitch of CMOS image sensors is comparable to that of the CCDs, shown as + signs in the figure.

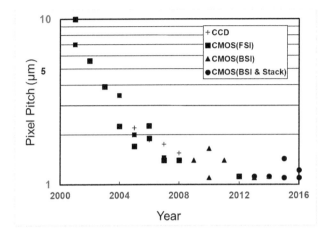

FIGURE 1.2: Trend of pixel pitch in 4T-APS type CMOS imagers. The solid squares and + signs show the pixel pitch for CMOS imagers and CCDs, respectively.

1.3   Brief history of smart CMOS image sensors

1.3.1   Vision chips

There are three main categories of smart CMOS image sensors, namely pixel-level processing, chip-level processing or camera-on-a-chip, and column-level processing as shown in Fig. 1.3. The first category, i.e., the pixel-parallel processing, is also called the vision chip. In the 1980s, C. Mead and coworkers at Caltech* proposed and demonstrated vision chips or silicon retina [44]. A silicon retina mimics the human visual processing system with a massively parallel-processing capability using the Si LSI technology. The circuits work in the sub-threshold region, as discussed in Appendix F, to achieve low power consumption. In addition, the circuits automatically solve a given problem by using convergence in 2D resistive networks [44]. They frequently use photo-transistors (PTrs) as photo-detectors owing to the gain of PTrs. Since the 1980s, considerable work has been done on developing vision chips and similar devices, as reviewed by Koch and Liu [45], and A. Moini [1].

---

*California Institute of Technology.

Massively parallel processing in the focal plane is very attractive and has been the subject of much research in fields, such as programmable artificial retinas [46]. Some applications have been commercialized, such as two-layered resistive networks using 3T-APS by T. Yagi, S. Kameda, and coworkers at Osaka University [47, 48].

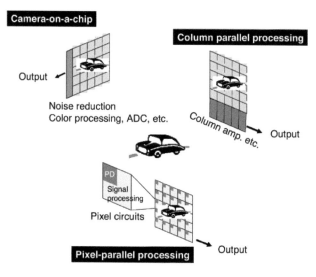

FIGURE 1.3: Three types of smart sensors.

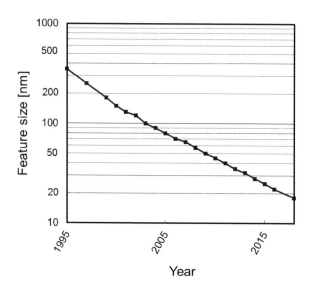

FIGURE 1.4: ITRS roadmap – the trend of DRAM half pitch [49].

Figure 1.4 shows the LSI roadmap of ITRS[†] [49]. This figure shows the trend of dynamic random access memory (DRAM) half pitch. Other technologies such as logic processes exhibit almost the same trend, namely the integration density of the LSI increases according to Moore's law such that it doubles every 18–24 months [50].

This advancement of CMOS technology means that massively parallel processing or pixel-parallel processing is becoming more feasible. Considerable research has been published, for example, on vision chips based on cellular neural networks (CNN) [51–54], programmable multiple instruction and multiple data (MIMD) vision chips [55], biomorphic digital vision chips [56], and analog vision chips [57, 58]. Other pioneering works include digital vision chips using pixel-level processing based on single instruction and multiple data (SIMD) processor by M. Ishikawa *et al.* at the University of Tokyo and Hamamatsu Photonics [59–65].

It should be noted that some vision chips are not based on the human visual processing system, and thus they belong in the category of pixel-level processing.

1.3.2   Advancement of CMOS technology and smart CMOS image sensors

The second category, i.e., chip-level processing, is more tightly related with the advancement of CMOS technology and has little relation with the pixel-parallel processing. This category includes system-on-chip and system-on-a-camera. In the early 1990s, advancement of CMOS technology made it possible to realize highly integrated CMOS image sensors or smart CMOS image sensors for machine vision. Pioneering works include ASIC vision (originally developed in University of Edinburgh [66, 67] and later by VLSI Vision Ltd. (VVL)), near-sensor image processing (NSIP) (later known as PASIC [68]) originally developed in Linköping University [69] and MAPP by Integrated Vision Products (IVP) [70].

PASIC may be one of the first CMOS image sensors that uses a column level analog-to-digital converter (ADC) [68]. ASIC vision has a PPS structure [69], while NSIP uses a pulse width modulation (PWM) based sensor [69], which will be discussed in Sec. 3.2.2.1. On the other hand, MAPP uses an APS [70].

1.3.3   Smart CMOS image sensors based on high performance CMOS image sensor technologies

Some of the above mentioned sensors have column-parallel processing structure, which is the third of the categories. Column-parallel processing is suitable for CMOS image sensors, because the column lines are electrically independent of each other. Column-parallel processing can enhance the performance of CMOS image sensors, for example, by widening the dynamic range and increasing the speed. Combined with the 4T-APSs, column-parallel processing exhibits high image quality

---

[†]International Technology Roadmap for Semiconductors.

and versatile functions. Therefore, recently column-level processing architecture has been widely used for higher performance CMOS image sensors. The advancement of CMOS technologies also broadens the range of applications of this architecture.

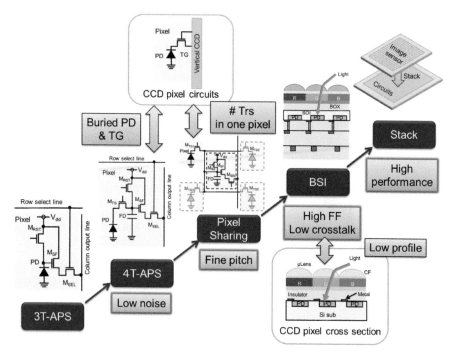

FIGURE 1.5: Key technologies for the advancement of CMOS image sensors. 3T-APS and 4T-APS are described in Sec. 2.5. The difference between CCDs and CMOS image sensors is discussed in Sec. 2.10. The pixel sharing techniques are introduced in Sec. 2.5.3.2. The backside illumination (BSI) and stacked technologies are mentioned in Sec. 3.3.3 and Sec. 3.3.4.

In 2010s, backside illumination (BSI) structure was introduced in commercial products, especially in smart phone cameras. Because the pixel pitch of the image sensors in smart phone cameras shrinks, the optical crosstalk increases, while the SNR degrades. The BSI structure, which is described in Sec. 3.3.3, can alleviate the above two issues, namely reducing the optical crosstalk and retaining or increasing the SNR even if the pixel count decreases. Its structure has been widely used and it has pushed the development of stacked CMOS image sensors. The opposite side of the BSI CMOS image sensor is a circuit layer, which can be attached to the other circuit layer of the other chip or the wafer. Thus, a stacked structure has emerged so as to integrate the signal processing functions in one chip. This suggests that the stacked CMOS image sensors can realize smart CMOS imaging if each pixel can be directly connected to the other signal processing layer. The stacked structures are mentioned in Sec. 3.3.4.

Advancement of CMOS technologies has improved basic performance of image sensors along the axes of ***Light Intensity*** and ***Time*** as shown in Fig. 1.6. This advancement can also be used to enhance functions realized by smart CMOS image sensors. This shows the third axis of *X* in Fig. 1.6. These applications shown in the figure are introduced in Chapters 3, 4, and 5.

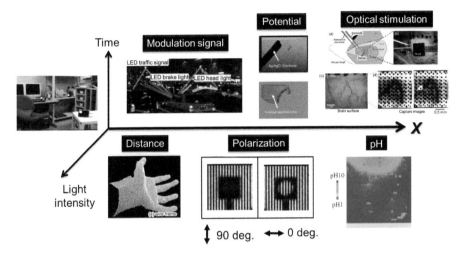

FIGURE 1.6: Applications of smart CMOS image sensors. The photos in this figure are taken from Fig. 5.2 for Modulation signal; Fig. 5.37 for Potential; Fig. 5.54 for Optical stimulation; Fig. 4.30 for Distance; Fig. 3.42 for Polarization and Fig. 5.43 for pH.

1.4 Organization of the book

This book is organized as follows. Firstly, in this introductory chapter, a general overview of solid-state image sensors is presented. This is followed by a description of smart CMOS image sensors, their brief history and salient features. Next, in Chapter 2, fundamentals of CMOS image sensors are presented in detail. Going into more details, the optoelectronic properties of silicon semiconductors, based on the CMOS technology, are described in Sec. 2.2. Then, in Sec. 2.3, several types of photo-detectors are introduced, including the photodiode, which is commonly used in CMOS image sensors. The operating principle and fundamental characteristics of photodiodes are described. In a CMOS image sensor, a photodiode is used in the accumulation mode, which is very different from the mode of operation for other applications such as optical communication. This accumulation mode is discussed in Sec. 2.4. The pixel structure is the heart of this chapter and is explained in Sec. 2.5 which covers both active pixel sensors (APS) and passive pixel sensors (PPS). The peripheral blocks, other than pixels, are described in Sec. 2.6. Addressing and readout circuits are also mentioned in that section. The fundamental characteristics of CMOS image sensors are discussed in Sec. 2.7. The topics of color and pixel sharing are also described in Sec. 2.8 and Sec. 2.5.3.2, respectively. Finally, several comparisons are discussed in Sec. 2.9 and Sec. 2.10.

In Chapter 3, several smart functions and materials are introduced. Certain smart CMOS image sensors have been developed by introducing new functions in the conventional CMOS image sensor architecture. Firstly, pixel structures (e.g., the log sensor) that are different from those of conventional APS are introduced in Sec. 3.2. Smart CMOS image sensors can be classified into three categories: analog, digital, and pulse, and these are described in Sections 3.2.1, 3.2.2, and 3.2.3, respectively. The CMOS image sensors are typically based on silicon CMOS technologies; however, other technologies and materials can be used to achieve smart functions. For example, silicon on sapphire (SOS) technology is a candidate for smart CMOS image sensors. Section 3.3.2.2 discusses materials other than Si in smart CMOS image sensors. Structures other than the standard CMOS technologies are described in Sec. 3.3. Sec. 3.4 considers two types of smart CMOS image sensors with nonorthogonal pixel arrangements and dedicated optics.

By combining the smart functions introduced in Chapter 3, Chapter 4 describes several examples of smart imaging. High sensitivity (Sec. 4.2), high speed (Sec. 4.3), and wide dynamic range (Sec. 4.4) are presented with examples. These features of smart CMOS image sensors give a higher performance compared to the conventional CMOS image sensors. Another feature of smart CMOS image sensors is that they can achieve versatile functions that cannot be realized by conventional image sensors. To discuss them, sections on demodulation (Sec. 4.5) and 3D range finders (Sec. 4.6) are presented.

The final chapter, i.e., Chapter 5, considers applications using the smart CMOS image sensors in the field of information and communication technologies, biotechnologies, and medicine. These applications have recently emerged and will be important for the next generation of smart CMOS image sensors.

Several appendices are attached at the end of this book to supplement the information provided in the main body of the book.

2

Fundamentals of CMOS image sensors

2.1   Introduction

This chapter provides the fundamental knowledge required for understanding CMOS image sensors. A CMOS image sensor generally consists of an imaging area, which comprises an array of pixels, vertical and horizontal access circuitry, and readout circuitry, as shown in Fig. 2.1.

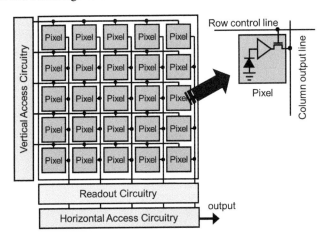

FIGURE 2.1: Architecture of a CMOS image sensor. A 2D array of pixels, vertical and horizontal access circuitry, and readout circuitry are generally implemented. A pixel consists of a photo-detector and transistors.

The imaging area is a 2D array of pixels, with each pixel containing a photo-detector and some transistors. This area is the heart of the image sensor and the imaging quality is largely determined by the performance of this area. The access circuitry is used to access a pixel and read the signal value in the pixel. Usually a scanner or shift register is used for the purpose, and a decoder is used to access the pixels randomly, which is sometimes important for smart sensors. A readout circuit is a 1D array of switches and a sample and hold (S/H) circuit. Noise canceling circuits, such as correlated double sampling (CDS), are employed in this area.

In this chapter, the fundamental elements of CMOS image sensors are described.

Firstly, the concept of photo-detection is explained. The behavior of minority carriers plays an important role in photo-detection. Several kinds of photo-detectors for CMOS image sensors are introduced. Among them, the pn-junction photodiodes are used, the most often. Hence, the operating principle and the basic characteristics of pn-junction photodiodes are explained in detail. In addition, the accumulation mode, which is an important operation for the CMOS image sensors, is described. Then, some basic pixel structures are introduced, namely passive pixel sensors (PPSs) and active pixel sensors (APSs). Finally, further elements of CMOS image sensors are described, such as scanners and decoders, read-out circuits, and noise cancelers.

2.2 Fundamentals of photo-detection

2.2.1 Absorption coefficient

When light is incident on a semiconductor, a part of the incident light is reflected while the rest is absorbed in the semiconductor and produces electron–hole pairs inside the semiconductor, as shown in Fig. 2.2. Such electron–hole pairs are called photo-generated carriers. The number of photo-generated carriers depends on the semiconductor material and is described by the absorption coefficient α.

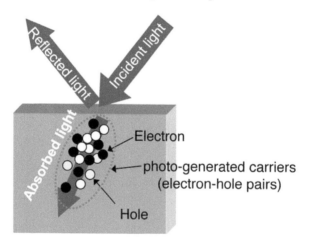

FIGURE 2.2: Photo-generated carriers in a semiconductor.

It should be noted that α is defined as the fractional decrease of light power $\Delta P/P$, as light travels a distance Δz; that is,

$$\alpha(\lambda) = \frac{1}{\Delta z} \frac{\Delta P}{P}. \tag{2.1}$$

From Eq. 2.1, the following relation is derived:

$$P(z) = P_o \exp(-\alpha z).$$ (2.2)

The absorption length L_{abs} is defined as

$$L_{abs} = \alpha^{-1}.$$ (2.3)

It should be noted that α is a function of the photon energy $h\nu$ or wavelength λ, where h and ν are Planck's constant and the frequency of light, respectively. The value of L_{abs}, thus depends on λ. Figure 2.3 shows the dependence of α and L_{abs} of Si on λ of the input light. In the visible region, i.e., 0.4–0.6 μm wavelength region, L_{abs}t lies within 0.1–10 μm [71]. L_{abs} is an important parameter used in the approximate estimation of a photodiode structure.

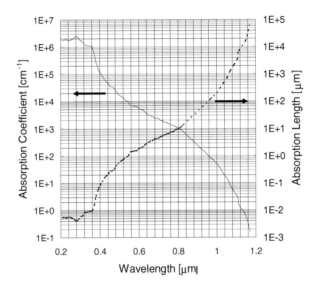

FIGURE 2.3: Absorption coefficient (solid line) and absorption length (broken line) of silicon as a function of wavelength. From the data in [71].

2.2.2 Behavior of minority carriers

The light incident on a semiconductor generates electron–hole pairs or photo-generated carriers. When electrons are generated in a p-type region, the electrons are minority carriers. The behavior of minority carriers is important for image sensors. For example, in a CMOS image sensor with a p-type substrate, photo-generated minority carriers in the substrate are electrons. This situation occurs when near infrared (NIR) light is incident on the sensor, because L_{abs} in the NIR region is over 1 μm, as shown in Fig. 2.3, and thus the light reaches the substrate.

In that case, the diffusion behavior of the carriers greatly affects the image sensor characteristics. In other words, the carriers can diffuse to adjacent photodiodes through the substrate and cause image blurring. To suppress this, an NIR cut filter is usually used, because the NIR light reaches deeper regions of the photodiode, namely the substrate, and produces more carriers than does the visible light.

The mobility and lifetime of minority carriers are empirically given by the following relations [72–74] with parameters of acceptor concentration N_a and donor concentration N_d:

$$\mu_n = 233 + \frac{1180}{1 + [N_a/(8 \times 10^{16})]^{0.9}} \ [\text{cm}^2/\text{V} \cdot \text{s}], \tag{2.4}$$

$$\mu_p = 130 + \frac{370}{1 + [N_d/(8 \times 10^{17})]^{1.25}} \ [\text{cm}^2/\text{V} \cdot \text{s}], \tag{2.5}$$

$$\tau_n^{-1} = 3.45 \times 10^{-12} N_a + 0.95 \times 10^{-31} N_a^2 \ [\text{s}^{-1}], \tag{2.6}$$

$$\tau_p^{-1} = 7.8 \times 10^{-13} N_a + 1.8 \times 10^{-31} N_d^2 \ [\text{s}^{-1}]. \tag{2.7}$$

From the above equations, we can estimate the diffusion lengths $L_{n,p}$ for electrons and holes by using the relation

$$L_{n,p} = \sqrt{\frac{k_B T \mu_{n,p} \tau_{n,p}}{e}}. \tag{2.8}$$

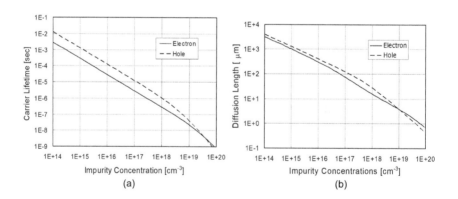

FIGURE 2.4: Lifetimes (a) and diffusion lengths (b) of electrons and holes in silicon as a function of impurity concentration.

Figure 2.4 shows the lifetimes and diffusion lengths of electrons and holes as functions of impurity concentration. Note that both electrons and holes can travel over 100 μm for impurity concentrations below 10^{17} cm^{-3}.

2.2.3 Sensitivity and quantum efficiency

The sensitivity (photo-sensitivity) R_{ph} is defined as the amount of photo-current I_L produced, when a unit of light power P_o is incident on a material. It is given by

$$R_{ph} \equiv \frac{I_L}{P_o}. \tag{2.9}$$

The quantum efficiency η_Q is defined as the ratio of the number of generated photo-carriers to the number of the input photons. The input photon number per unit time and the number of generated carrier per unit time are I_L/e and $P_o/(h\nu)$, respectively, and hence η_Q is expressed as

$$\eta_Q \equiv \frac{I_L/e}{P_o/(h\nu)} = R_{ph}\frac{h\nu}{e}. \tag{2.10}$$

From Eq. 2.10, the maximum sensitivity, that is R_{ph} at $\eta_Q = 1$, is found to be

$$R_{ph,max} = \frac{e}{h\nu} = \frac{e}{hc}\lambda = \frac{\lambda \; [\mu m]}{1.23}. \tag{2.11}$$

$R_{ph,max}$ is illustrated in Fig. 2.5. It monotonically increases in proportion to the wavelength of the input light and eventually reaches zero at a wavelength λ_g corresponding to the bandgap of the material E_g. For Si, the wavelength is approximately $1.12 \; \mu m$, as the bandgap of Si is $1.107 \; eV$.

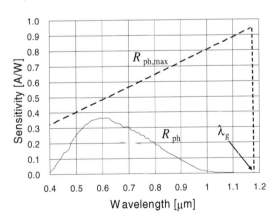

FIGURE 2.5: Sensitivity of silicon. The solid line shows the sensitivity (R_{ph}) according to Eq. 2.19. The dashed line shows the ideal sensitivity or maximum sensitivity ($R_{ph,max}$) according to Eq. 2.11. λ_g is the wavelength at the bandgap of silicon.

2.3   Photo-detectors for smart CMOS image sensors

Most photo-detectors used in CMOS image sensors are pn-junction photodiodes (PDs). In the following sections, PDs are described in detail. Other photo-detectors used in CMOS image sensors are photo-gates (PGs), photo-transistors (PTrs), and avalanche photodiodes (APDs). PTrs and APDs both make use of gain. Another detector with gain is the photo-conductive detector (PCD). Figure 2.6 illustrates the structures of PDs, PGs, and PTrs.

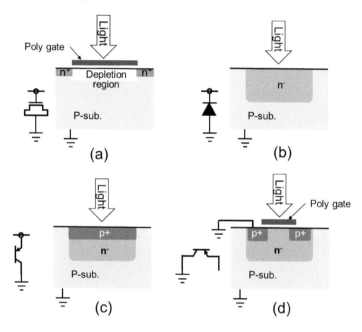

FIGURE 2.6: Symbols and structures of (a) photodiode, (b) photo-gate, (c) vertical type photo-transistor, and (d) lateral type photo-transistor.

2.3.1   pn-junction photodiode

In this section, a conventional photo-detector, namely a pn-junction PD is described [75, 76]. First, the operating principle of a PD is described and then, several fundamental characteristics, such as quantum efficiency, sensitivity, dark current, noise, surface recombination, and speed, are discussed. These characteristics are important for smart CMOS image sensors.

2.3.1.1   Operating principle

The operating principle of a pn-junction PD is quite simple. In a pn-junction diode, the forward current I_F is expressed as

$$I_F = I_{diff} \left[\exp \left(\frac{eV}{nk_BT} \right) - 1 \right], \qquad (2.12)$$

where n is the ideal factor and I_{diff} is the saturation current or diffusion current, which is given by

$$I_{diff} = eA \left(\frac{D_n}{L_n} n_{po} + \frac{D_p}{L_p} p_{no} \right), \qquad (2.13)$$

where $D_{n,p}$, $L_{n,p}$, n_{po}, and p_{no} are the diffusion coefficient, diffusion length, minority carrier concentration in the p-type region, and the minority carrier concentration in n-type region, respectively. A is the cross-sectional area of the pn-diode. The photo-current of the pn-junction photodiode is expressed as follows:

$$\begin{aligned} I_L &= I_{ph} - I_F \\ &= I_{ph} - I_{diff} \left[\exp \left(\frac{eV}{nk_BT} \right) - 1 \right], \end{aligned} \qquad (2.14)$$

where n is the ideal factor.

Figure 2.7 illustrates the I–V curves of a pn-PD under dark and illuminated conditions. There are three modes for bias conditions, namely solar cell mode, PD mode, and avalanche mode, as shown in Fig. 2.7.

Solar cell mode   In the solar cell mode, no bias is applied to the PD. Under light illumination, the PD acts as a battery; that is, it produces a voltage across the pn-junction. Figure 2.7 shows the open circuit voltage V_{oc}. In the open circuit condition, the voltage V_{oc} can be obtained by substituting $I_L = 0$ in Eq. 2.14, and thus

$$V_{oc} = \frac{k_BT}{e} \ln \left(\frac{I_{ph}}{I_{diff}} + 1 \right). \qquad (2.15)$$

This shows that the open circuit voltage does not linearly increase with the input light intensity.

PD mode   The second mode is the PD mode. When a PD is reverse biased, i.e., $V < 0$, the exponential term in Eq. 2.14 can be neglected, and thus, I_L becomes

$$I_L \approx I_{ph} + I_{diff}. \qquad (2.16)$$

This shows that the output current of the PD is equal to the sum of the photo-current and the diffusion current. Thus, the photo-current linearly increases with the input light intensity.

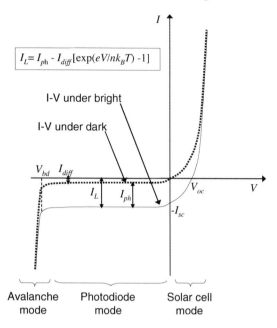

FIGURE 2.7: PD I–V curves under dark and bright conditions.

Avalanche mode   The third mode is the avalanche mode. When a PD is strongly biased, the photo-current suddenly increases, as shown in Fig. 2.7. This phenomenon is called an avalanche, wherein impact ionization of electrons and holes occurs, and the carriers are multiplied. The voltage, where an avalanche occurs is called the avalanche breakdown voltage V_{bd} shown in Fig. 2.7. Avalanche breakdown is explained in Sec. 2.3.1.3. The avalanche mode is used in an avalanche photodiode (APD) and is described in Sec. 2.3.4.

2.3.1.2   Quantum efficiency and sensitivity

By using the definition of α in Eq. 2.2, the light intensity is expressed as

$$dP(z) = -\alpha(\lambda)P_o \exp\left[-\alpha(\lambda)z\right] dz. \tag{2.17}$$

To emphasize that the absorption coefficient is dependent on the wavelength, α is written as $\alpha(\lambda)$. The quantum efficiency η_Q is defined as the ratio of the absorbed light intensity to the total input light intensity, and thus

$$\eta_Q = \frac{\int_{x_n}^{x_p} \alpha(\lambda)P_o \exp\left[-\alpha(\lambda)x\right] dx}{\int_0^\infty \alpha(\lambda)P_o \exp\left[-\alpha(\lambda)x\right] dx} \tag{2.18}$$
$$= (1 - \exp\left[-\alpha(\lambda)W\right]) \exp\left[-\alpha(\lambda)x_n\right],$$

where W is the depletion width and x_n is the distance from the surface to the edge of the depletion region as shown in Fig. 2.8.

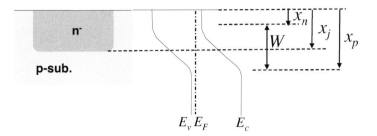

FIGURE 2.8: pn-junction structure. The junction is formed at a position x_j from the surface. The depletion region widens at the sides of the n-type region x_n and p-type region x_p. The width of the depletion region W is thus equal to $x_n - x_p$.

Using Eq. 2.18, the sensitivity R_{ph} is expressed as follows:

$$R_{ph} = \eta_Q \frac{e\lambda}{hc}$$
$$= \frac{e\lambda}{hc}(1 - \exp[-\alpha(\lambda)W])\exp[-\alpha(\lambda)x_n]. \tag{2.19}$$

In this equation, the depletion width W and the part of the depletion width at the n– region x_n are expressed as follows. Using the built-in potential V_{bi}, W under an applied voltage V_{appl} is expressed as

$$W = \sqrt{\frac{2\varepsilon_{Si}(N_d + N_a)(V_{bi} + V_{appl})}{eN_aN_d}}, \tag{2.20}$$

where ε_{Si} is the dielectric constant of silicon. The built-in potential V_{bi} is given by

$$V_{bi} = k_B T \ln\left(\frac{N_dN_a}{n_i^2}\right), \tag{2.21}$$

where n_i is the intrinsic carrier concentration for silicon and $n_i = 1.4 \times 10^{10}$ cm^{-3}. The parts of the depletion width at the n-region and p-region are

$$x_n = \frac{N_a}{N_a + N_d}W, \tag{2.22}$$

$$x_p = \frac{N_d}{N_a + N_d}W. \tag{2.23}$$

Figure 2.9 shows the sensitivity spectrum curve of silicon, that is, the dependence of the sensitivity on the input light wavelength. The sensitivity spectrum curve is dependent on the impurity profile of the n-type and p-type regions as well as the position of the pn-junction x_j. In the calculation of the curve in Fig. 2.9, the impurity profile in both the n-type and p-type regions is flat and the junction is abrupt. In addition, only the photo-generated carriers in the depletion region are accounted for. Some portion of the photo-generated carriers outside the depletion region diffuses and reaches the depletion region, but is unaccounted for in the calculation of these diffusion carriers. Such diffusion carriers can affect the sensitivity at long wavelengths because of the low value of the absorption coefficient in the long wavelength region [77]. Another assumption is made here, which involves neglecting the surface recombination effect, which will be considered in the section on noise (see Sec. 2.3.1.4). These assumptions will be further discussed in Sec. 2.4. An actual PD in an image sensor is coated with SiO_2 and Si_3N_4, and thus the quantum efficiency is changed [78].

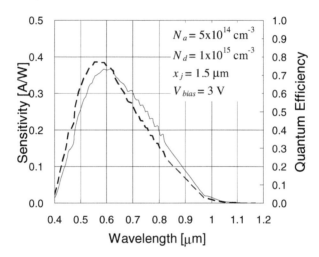

FIGURE 2.9: Dependence of the sensitivity (solid line) and quantum efficiency (broken line) of a pn-junction PD on wavelength. The PD parameters are summarized in the inset.

2.3.1.3   Dark current

Dark current in PDs has several sources, as listed below.

Diffusion current   The diffusion current inherently flows and is expressed as

$$
\begin{aligned}
I_{diff} &= Ae\left(\frac{D_n n_{po}}{L_n} + \frac{D_p n_{no}}{L_p}\right) \\
&= Ae\left(\frac{D_n}{L_n N_a} + \frac{D_p}{L_p N_d}\right) N_c N_v \exp\left(-\frac{E_g}{k_B T}\right),
\end{aligned}
\tag{2.24}
$$

where A is the diode area, N_c and N_v are the effective densities of the states in the conduction band and the valence band, respectively, and E_g is the bandgap. Thus, the diffusion current exponentially increases with the temperature. It should be noted that the diffusion current does not depend on the bias voltage.

Tunnel current   In addition to the above, there are other dark currents, which include tunnel current, generation–recombination (g–r current), Frakel–Poole current, and surface leak current [79,80]. The tunnel current consists of band-to-band tunneling (BTBT) current and trap-assisted tunneling (TAT), which have an exponential dependence on the bias voltage [79–81]. However, they have only a little dependence on the temperature. Although both BTBT and TAT cause dark current to be exponentially dependent on the bias voltage, the dependence is different, as shown in Table 2.1. The tunnel current is important when doping is large and thus the depletion width becomes thin so as to lead to tunneling.

G–R current   In the depletion region, the carrier concentration is reduced and carrier generation occurs rather than the recombination of carriers [79,82] by thermal generation. This gives rise to dark current. The G–R current is given by [79]

$$I_{gr} = AW \frac{en_i}{\tau_g} = AW \frac{e\sqrt{N_c N_v}}{\tau_g} \exp\left(-\frac{E_g}{2k_B T}\right),\tag{2.25}$$

where W is the depletion width, τ_g is the lifetime of the deep level, and n_i is the intrinsic carrier concentration. Since W is proportional to $\sqrt{V}$, I_{gr} is also proportional to $\sqrt{V}$. This process is called Shockley–Read–Hall (SRH) recombination [79,82].

Impact ionization current   Impact ionization or avalanche breakdown increases the dark current when the bias voltage increases [83, 84]. The bias dependence of the dark current by impact ionization arises from the voltage dependence of the ionization coefficients of electrons and holes, α_n and α_p, respectively. These coefficients exponentially increase as the bias voltage increases.

Frankel–Poole current   The Frankel–Poole current originates from the emission of trapped electrons into the conduction band [79]. This current strongly depends on the bias voltage, which is same as the tunneling current.

Surface leak current   The surface leak current is given by

$$I_{surf} = \frac{1}{2} e n_i s_o A_s,\tag{2.26}$$

where n_i, s_o, and A_s are the intrinsic carrier concentration, surface recombination rate, and surface area, respectively.

TABLE 2.1
Dependence of dark current on temperature and voltage [79]. a, a', b, c, and d are constants; ϕ_B is a barrier height.

Process	Dependence
Diffusion	$\propto \exp\left(-\frac{E_g}{k_B T}\right)$
G–R	$\propto \sqrt{V} \exp\left(-\frac{E_g}{2k_B T}\right)$
Band-to-band tunneling	$\propto V^2 \exp\left(-\frac{a}{V}\right)$
Trap-assisted tunneling	$\propto \exp\left(-\frac{a'}{V}\right)^2$
Impact ionization	$\propto \exp\left(-\frac{b}{V}\right)$
Frankel–Poole	$\propto V \exp\left(-\frac{e}{k_B T}\left(d\sqrt{V} - \phi_B\right)\right)$
Surface leak	$\propto \exp\left(-\frac{E_g}{2k_B T}\right)$

Dependence of dark current on temperature and bias voltage   A comparison of Eqs. 2.24, 2.25, and 2.26 shows that the temperature dependences of various dark currents are different; the band-to-band tunneling current is independent of temperature, while $\log I_{diff}$ and $\log I_{gr}$ vary as $-\frac{1}{T}$ and $-\frac{1}{2T}$, respectively. Thus, the temperature dependence can reveal the origin of the dark current. Furthermore, their dependence on the bias voltage is different. The dependence of various dark currents on temperature and bias voltage is summarized in Table 2.1.

2.3.1.4   Noise

In this section, several types of noise in PD are described. The other noises that are inherent to CMOS image sensors, such as fixed pattern noise (FPN), are discussed in Sec. 2.7.1.

Shot noise   A PD suffers from shot noise and thermal noise. The shot noise originates from the fluctuations in the number of the particles N per second (for example, electrons and photons). Thus, photon shot noise and electron (or hole) shot noise inherently exist in a PD. The photon shot noise is produced by the incident light, while the electron shot noise is caused by a dark current. Photon or electron flux is a stochastic process and follows the Poisson distribution. In the Poisson distribution, the variation is given as $\sigma_N = \sqrt{N}$, where N is the value of photon or electron flux (particle numbers per second). Thus, the signal-to-noise ratio (SNR) is expressed as

$$\text{SNR} = \frac{N}{\sqrt{N}} = \sqrt{N}. \tag{2.27}$$

The electron shot noise is expressed in terms of shot noise current and the root mean square of the shot noise current i_{sh} is expressed as

$$i_{sh,\,rms} = \sqrt{2e\bar{I}\Delta f}, \tag{2.28}$$

where $\bar{I}$ and Δf indicate average signal current and bandwidth, respectively. The SNR for shot noise is expressed as

$$\text{SNR} = \frac{\bar{I}}{\sqrt{2e\bar{I}\Delta f}} = \frac{\sqrt{I}}{2e\Delta f}. \tag{2.29}$$

Thus, as the amount of current or the number of electrons decreases, the SNR associated with the shot noise decreases. In PDs, the electron shot noise is mainly produced by the dark current, and thus, does not depend on the input light intensity, L. On the other hand, SNR determined by photon shot noise is proportional to $\sqrt{L}$. Thus, in the low light region, the electron shot noise limits S/N, while, in the bright light region, the photon shot noise limits it.

Thermal noise   In a load resistance R, free electrons exist and randomly move according to the temperature of the load resistance. This effect generates thermal noise, also known as Johnson noise or Nyquist noise. It is also called white noise, because the thermal noise spreads over the frequency. The thermal noise is expressed as

$$i_{sh,\,rms} = \sqrt{\frac{4k_B T \Delta f}{R}}. \tag{2.30}$$

In CMOS image sensors, the thermal noise appears as $k_B TC$ noise, which is discussed in Sec. 2.7.1.2.

2.3.1.5 Surface recombination

In a conventional CMOS image sensor, the surface of the silicon is interfaced with SiO_2 and has some dangling bonds, which produce surface states or interface states acting as non-recombination centers. Some photo-generated carriers near the surface are trapped at the centers and do not contribute to the photo-current. Thus these surface states degrade the quantum efficiency or sensitivity. This effect is called surface recombination. The feature parameter for the surface recombination is the surface recombination rate S_{surf}, and is proportional to the excess carrier density at the surface:

$$D_n \frac{\partial n_p}{\partial x} = S_{surf}\left[n_p(0) - n_{po}\right]. \tag{2.31}$$

S_{surf} is strongly dependent on the interface state, band bending, defects, and other effects, and is approximately 10 cm³/s for both electrons and holes. For short wavelengths, such as blue light, the absorption coefficient is large and the absorption mostly occurs at the surface. Thus, it is important to reduce the surface recombination velocity to achieve high quantum efficiency in the short wavelength region.

2.3.1.6   Speed

In the recent growth of optical fiber communications and fiber-to-the-home (FTTH) technology, silicon CMOS photo-receivers have been studied and developed. High-speed photo-detectors using CMOS technologies, including BiCMOS (Bipolar CMOS) technology, are described in detail in Ref. [85, 86], and high-speed circuits for CMOS optical fiber communication are also detailed in Ref. [87].

Here we consider the response of a PD, which is generally limited by the CR time constant τ_{CR}, transit time τ_{tr}, and diffusion time of the minority carriers τ_n for the electrons:

- The CR time originates from the pn-junction capacitance C_D and is expressed as

$$\tau_{CR} = 2\pi C_D R_L, \tag{2.32}$$

  where R_L is the load resistance.

- The transit time across the depletion region here is defined as the time for a carrier to drift across the depletion region. It is expressed as

$$\tau_{tr_depletion} = W/v_s, \tag{2.33}$$

  where v_s is the saturation velocity.

- The minority carriers generated outside the depletion region can reach the depletion region after the diffusion time,

$$\tau_{n,p} = L_{n,p}^2/D_{n,p}, \tag{2.34}$$

  for electrons with a diffusion coefficient of D_n.

It should be noted there is a trade-off between the depletion width W and the quantum efficiency η_Q in the case of a transit time limitation. In this case,

$$\eta_Q = [1 - \exp(-\alpha(\lambda)v_s t_{tr})]\exp(-\alpha(\lambda)x_n). \tag{2.35}$$

Of the above, the diffusion time has the greatest effect on the PD response in CMOS image sensors.

In conventional image sensors, the response speed of a PD is not a concern. However, some types of smart image sensors need a PD with a fast response. Smart CMOS image sensors for optical wireless LANs are an example and are considered in Chapter 5. They are based on technologies for CMOS-based photo-receivers for optical fiber communications, mentioned above.

Another example is a smart CMOS image sensor that can measure the time-of-flight (TOF), which is also described in Chapter 5. In this case, APDs and other fast response PDs are used for their high-speed response. It should be noted that a high speed PD, such as an APD has a high electric field in the vertical direction, and almost no electric field in the horizontal direction. In 4T-APS type CMOS image sensors, which are mentioned in Sec. 2.5.3, the photo-generated carriers diffuse so as to be transferred to the next node, and thus a drift mechanism is introduced for ultra high-speed response in PD. This is discussed in Sec. 2.7.3 and its application to high-speed CMOS image sensors is presented in Section 4.3.4 of Chapter 4.

2.3.2 Photo-gate

The structure of a photo-gate (PG) is the same as that of a MOS capacitor. The photo-generated carriers accumulate in the depletion region when the gate is biased. A PG has a suitable structure to accumulate and transfer the carriers, and has been used in some CMOS image sensors. The accumulation of photo-generated carriers in a PG is shown in Fig. 2.10. By applying a gate bias voltage, a depletion region is produced, which acts as an accumulation region for photo-generated carriers, as shown in Fig. 2.10.

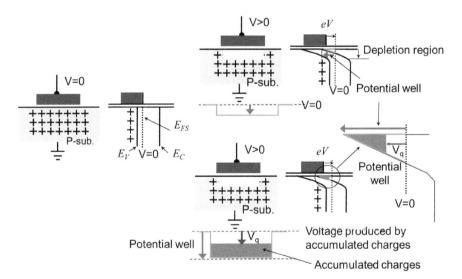

FIGURE 2.10: Photogate structure with applied gate voltage, which produces a depletion region where photo-generated carriers accumulate.

The fact that the photo-generated area is separated from the top surface in a PG is useful for some smart CMOS image sensors, as will be discussed in Chapter 5.

It should be noted that PGs have certain disadvantages with regard to sensitivity, because the gate, which is usually made of poly-silicon, is partially transparent and has an especially low transmittance at shorter wavelength or in the blue light region.

2.3.3   Photo-transistor

A photo-transistor (PTr) can be made using the standard CMOS technology for a parasitic transistor. A PTr amplifies the photo-current by a factor of the base current gain β. Because the base width and carrier concentration are not optimized by the standard CMOS technology, β is not high, typically about 10–20. In particular, the base width is a trade-off factor for the photo-transistor; when the base width increases, the quantum efficiency increases but the gain decreases [88]. Another disadvantage of a PTr is the large variation of β, which produces a fixed pattern noise (FPN), as detailed in Sec. 2.7.1.1. In spite of these disadvantages, PTrs are used in some CMOS image sensors owing to their simple structure and gain. When accompanied by current mirror circuits, PTrs can be used in current-mode signal processing, as discussed in Sec. 3.2.1.1. To address the low β at a low photo-current, a vertical inversion-layer emitter pnp BJT structure has been developed [89].

2.3.4   Avalanche photodiode

An avalanche photodiode (APD) utilizes an avalanche effect in which photo-generated carriers are multiplied [88]. APDs have considerable gain, as well as high-speed response. APDs are thus used as high speed detectors in optical fiber communication and ultra-low-light detection applications, such as biotechnology. However, they are hardly used in image sensors for the following reasons. Firstly, they require a high voltage close to the breakdown voltage. Such a high voltage hinders the use of APDs in the standard CMOS technologies, other than in hybrid image sensors with other APD materials with a CMOS readout circuit substrate, for example, as reported in Ref. [90]. Secondly, even if such a high voltage is acceptable, the gain is an analog value with large variation, and hence, the performance may be deteriorated. The gain variation causes the same problem as seen in the PTrs.

Pioneering work by A. Biber *et al.* at Centre Suisse d'Electronique et de Microtechnique (CSEM) has produced a 12×24-pixel APD array fabricated in a standard 1.2-μm BiCMOS technology [91]. Each pixel employs an APD control and readout circuits. An image is obtained with the fabricated sensor with an avalanche gain of about 7 under a bias voltage of 19.1 V.

2.3.4.1   Geiger mode APD

Several reports have been published of APDs fabricated using the standard CMOS technologies [92–95], as shown in Fig. 2.11.

In these reports, the APD is biased over the avalanche breakdown voltage, and thus when the photons are incident on the APD, it quickly turns on. As shown in Fig. 2.12, by being connected serially with a quench resistor, it can be turned off

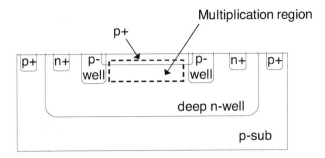

FIGURE 2.11: Cross section of the structure of SPAD. Illustrated after [93].

and thus, produces a spike-like current pulse as shown in Fig. 2.12. It can detect a single photon and hence, it is called single photon avalanche diode (SPAD). The initial position of SPAD is shown in (1) in Fig. 2.12 (b). When a photon hits on a SPAD, it quickly turns on ((1) → (2)) because it is biased well above the avalanche breakdown voltage, V_o in Fig. 2.12 (b), and a photocurrent flows. This photocurrent drops the bias voltage to the SPAD by means of the quench resistor R_q. Thus, the applied voltage to the SPAD goes under the avalanche breakdown voltage, and finally stops the breakdown ((2) → (3)). After quenching, the SPAD is charged again and recovered at the initial point (1). This phenomenon resembles that of a Geiger counter and thus it is called the Geiger mode.

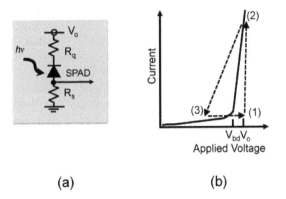

(a)   (b)

FIGURE 2.12: Operating principle of SPAD: (a) Basic circuits of SPAD with a quench resistor R_o, and (b) I-V curve of SPAD with a quench resistor. V_{bd} is avalanche breakdown voltage, and V_o is initial bias voltage to the SPAD. (1) initial point, (2) photocurrent flow, and (3) quench.

The difference between a conventional APD and a SPAD lies in its bias voltage [96]. APD is biased just around the avalanche breakdown voltage, V_{bd}, while SPAD is biased well above V_{bd}. Such bias voltage induces linear charge multiplication in the APD, and therefore, the output signal is proportional to the input light intensity.

On the contrary, in a SPAD, deep bias voltage well above V_{bd} triggers a diverging carrier multiplication so as to produce a large photocurrent.

Geiger-mode APD or SPAD can be fabricated in a standard CMOS technology. For example, in Ref. [97], shallow trench isolation (STI) was used as a guard ring for the APD. A bias voltage of 2.5 V is found to be sufficient to achieve avalanche breakdown. This result suggests a CMOS image sensor with a SPAD pixel array, which is described in Sec. 3.2.2.3 of Chapter 3. SPADs can be used in 3D range finder as described in Sec. 4.6.2 of Chapter 4, and biomedical applications such as fluorescence lifetime imaging microscopy (FLIM) as described in Sec. 5.4 of Chapter 5.

2.3.4.2   VAPD: Vertical APD

Recently, an APD suitable for photo-sensors for image sensors has been developed, which is called VAPD (vertical APD) [98, 99]. VAPD is fabricated in a backside illumination (BSI) CMOS image sensor, which is mentioned in Sec. 3.3.3 of Chapter 3 . Thanks to the BSI structure, it is possible to incorporate a vertical electric field in the photo-detection region. This structure also increases the FF (fill factor) compared with SPAD. In VAPD, a large gain saturates the output signal and thus VAPD image sensors produce a digital-like signal. By introducing multiple readout and accumulation of the output signal, high sensitivity characteristics can be achieved [99].

2.3.5   Photo-conductive detector

Another gain detector is the photo-conductive detector (PCD), which uses the effect of photo-conductivity [88]. A PCD typically has a structure of n^+–n^-–n^+. A DC bias is applied between the two n^+ sides and the generated electric field is largely confined to the n^- region, which is a photo-conductive region where electron–hole pairs are generated. The gain originates from a large ratio of the long lifetime of holes τ_p to the short transit time of electrons t_{tr}, that is, $\tau_p \gg t_{tr}$. The gain G_{PC} is expressed as

$$G_{PC} = \frac{\tau_p}{t_{tr}} \left(1 + \frac{\mu_p}{\mu_n} \right). \tag{2.36}$$

When a photo-generated electron–hole pair is separated by an externally applied electric field, the electron crosses the detector several times before recombining with the hole. It should be noted that a larger gain results in a slower response speed, that is, the gain-bandwidth is constant in a PCD, because G_{PC} is proportional to τ_p, which determines the response speed of the detector. Finally, a PCD has a relatively large dark current in general. As a PCD is essentially a conductive device, some dark current will flow. This may be a disadvantage for an image sensor. Some PCD materials are used as a detector overlaid on the CMOS readout circuitry in a pixel owing to the photo-response with a variety of wavelengths such as X-ray, UV, and infrared (IR). Avalanche phenomena occur in some PCD materials, realized in super

HARP, an imaging tube with ultra high sensitivity developed in NHK* [100].

Several types of CMOS readout circuitry (ROC) for this purpose have been reported, see Ref. [101] for example. Another application of PCD is as a replacement for on-chip color filters, described in Section 3.3.5 of Chapter 3 [102–104]. Some PDs are also used for fast photo-detectors and metal–semiconductor–metal (MSM) photo-detectors are used for this purpose.

Metal–semiconductor–metal photo-detector The MSM photo-detector is a kind of PCD, where a pair of metal fingers is placed on the surface of a semiconductor, as shown in Fig. 2.13 [88]. Because the MSM structure is easy to fabricate, MSM photo-detectors are also applied to other materials such as gallium arsenide (GaAs) and gallium nitride (GaN) . GaAs MSM photo-detectors are mainly used for ultra-fast photo-detectors [105], although in Refs. [106, 107] GaAs MSM photo-detector arrays are used for image sensors. GaN MSM photo-detectors have been developed for image sensors with sensitivity in UV region [108].

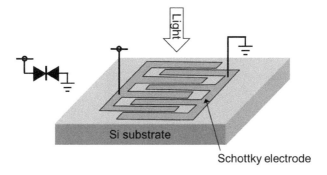

FIGURE 2.13: Structure of MSM photo-detector. The inset shows the symbols for the MSM photo-detector.

2.4   Accumulation mode in PDs

A PD in a CMOS image sensor is usually operated in the accumulation mode. In this mode, the PD is electrically floated and when light illuminates the PD, photocarriers are generated and swept to the surface owing to the potential well in the depletion region. The PD accumulation mode was proposed and demonstrated by G.P. Weckler [15]. The voltage decreases when electrons accumulate. By measuring the voltage

---

*Nippon Hoso Kyokai.

drop, the total amount of light power can be obtained. It should be noted that the accumulation of electrons is interpreted as the process of discharge in the charged capacitor by the generated photocurrent.

Let us consider a simple but typical case to understand why the accumulation mode is required in a CMOS image sensor. We assume the following parameters: the sensitivity of the PD R_{ph} = 0.6 A/W, the area size of the PD A = 100 μm^2, and the illumination at the PD surface L_o =100 lux. Assuming that 1 lux roughly corresponds to 1.6×10^{-7} W/cm^{-2} at λ=555 nm, as described in Appendix B, the photocurrent I_{ph} is evaluated as

$$
\begin{aligned}
I_{ph} &= R_{ph} \times L_o \times A \\
&= 0.6\,\text{A/W} \times 100 \times 1.6 \times 10^{-7}\,\text{W/cm} \times 100\,\mu\text{m}^2 \\
&\approx 10\,\text{pA}.
\end{aligned}
\tag{2.37}
$$

While it is possible to measure such a low photocurrent, it is difficult to precisely measure photocurrents of the same order from a 2D array for a large number of points at a video rate.

2.4.1   Potential change in accumulation mode

The junction capacitance of a pn-junction PD C_{PD} is expressed as

$$
C_{PD}(V) = \frac{\varepsilon_o \varepsilon_{Si}}{W},
\tag{2.38}
$$

which is dependent on the applied voltage V through the dependence of the depletion width W on V through the following relation:

$$
W = K(V + V_{bi})^{m_j},
\tag{2.39}
$$

where K is a constant, V_{bi} is the built-in potential of the p-n junction, and m_j is a parameter dependent on the junction shape: $m_j = 1/2$ and $1/3$ for a step and linear junctions, respectively.

The following will hold for $C_{PD}(V)$:

$$
C_{PD}(V)\frac{dV}{dt} + I_{ph} + I_d = 0,
\tag{2.40}
$$

where I_d is the dark current of the PD. Using Eqs. 2.38 and 2.39, Eq. 2.40 gives

$$
V(t) = (V_0 + V_{bi})\left[1 - \frac{(I_{ph} + I_d)(1 - m_j)}{C_0(V_0 + V_{bi})}t\right]^{\frac{1}{1-m_j}} - V_{bi},
\tag{2.41}
$$

where V_0 and C_0 are the initial values of the voltage and capacitance in the PD, respectively. This result shows that the voltage of the PD decreases almost linearly. Usually, the PD voltage is described as decreasing approximately linearly. Figure 2.14 shows the voltage drop of a PD as a function of time, and confirms that V_{PD} almost linearly decreases as time increases. Thus, light intensity can be estimated by measuring the voltage drop of the PD at a fixed time, usually at a video rate of 1/30 sec.

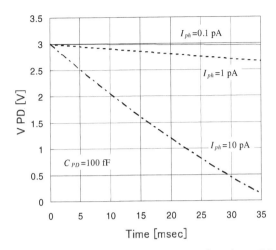

FIGURE 2.14: Voltage drop of a PD as a function of time.

2.4.2 Potential description

The potential description is frequently used for CMOS image sensors, and hence, it is an important concept. Figure 2.15 illustrates this concept [109]. In the figure, a MOSFET is depicted as an example; the source acts as a PD and the drain is connected to V_{dd}. The impurity density in the source is smaller than that in the drain. The gate is off-state, or in the subthreshold region.

Figure 2.15(b) shows the potential profile along the horizontal distance, showing the conduction band edge near the surface or the surface potential. In addition, the electron density at each area shown in Fig. 2.15 (c) is superimposed on the potential profile of (b); hence it is easy to see the carrier density profile, as shown in (d). The baseline of the carrier density profile sits at the bottom of the potential profile, and hence, the carrier density increases in the downward direction. It should be noted that the potential profile or Fermi level can be determined by the carrier density. When carriers are generated by input light and accumulate in the depletion region, the potential depth changes through the change in the carrier density. However, under ordinary conditions for image sensors, the surface potential increases in proportion to the accumulated charge.

Figure 2.16 shows the potential description of a PD, which is floated electrically. This is the same situation as in the previous section, Sec. 2.4.1. In the figure, the photo-generated carriers accumulate in the depletion region of the PD. The potential well V_b is produced by the built-in potential V_{bi} and the bias voltage V_j. Figure 2.16(b) illustrates the accumulated state when the photo-generated carriers are collected in the potential well. The accumulated charges change the potential depth from V_b to V_q, as shown in Fig. 2.16 (b). The amount of change $V_b - V_q$ is approximately proportional to the product of the input light intensity and accumulation time, as mentioned in Sec. 2.4.1.

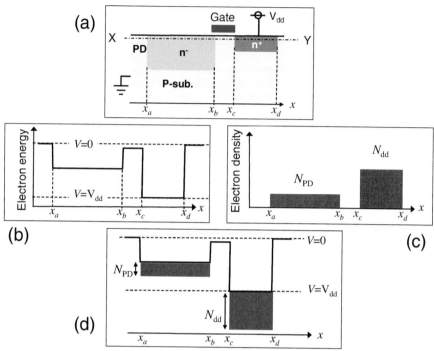

FIGURE 2.15: Illustration of potential description. An n-MOSFET structure is shown in (a), where the source is a PD and the drain is biased at V_{dd}. The gate of the MOSFET is off-state. The conduction band edge profile along X–Y in (a) is shown in (b). The horizontal axis shows the position corresponding to the position in (a) and the vertical axis shows the electron energy. $V = 0$ and $V = V_{dd}$ levels are shown in the figure. The electron density is shown in (c). (d) is the potential description, which is the superimposition of (a) and (c). Drawn after [109].

2.4.3 Behavior of photo-generated carriers in PD

As explained in Sec. 2.3.1.2, incident photons penetrate the semiconductor according to their energy or wavelength; photons with smaller energy or longer wavelength penetrate deeper into the semiconductor, while photons with larger energy or shorter wavelength are absorbed near the surface. The photons absorbed in the depletion region are swept immediately by the electric field and accumulate in the potential well, as shown in Fig. 2.17. In Fig. 2.17, lights of three colors, red, green, and blue, are incident on the PD. As shown in Fig. 2.17(a), the three lights reach different depths. The red light penetrates the deepest and reaches the p-substrate region, where it produces minority carrier electrons. In the p-type substrate region, there is little electric field, and hence, the photo-generated carriers only move by diffusion, as shown in Fig. 2.17(b). While some of the photo-generated carriers are recombined in this region and do not contribute to the signal charge, others arrive at the edge of the depletion region and accumulate in the potential well, thus contributing to the

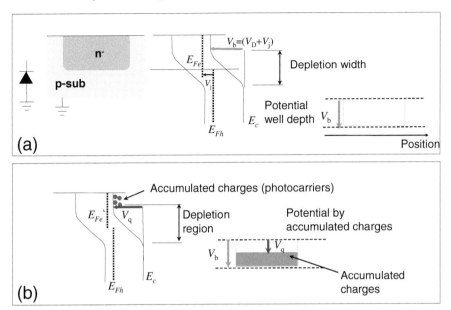

FIGURE 2.16: Potential description of a PD before accumulation (a) and after accumulation (b).

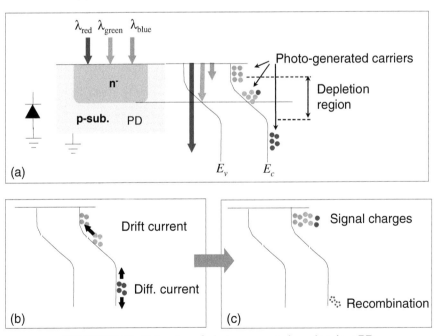

FIGURE 2.17: Behavior of photo-generated carriers in a PD.

signal charge. The extent of the contribution depends on the diffusion length of the carriers produced in the p-substrate, electrons in this case. The diffusion length has been discussed in Sec. 2.2.2. It should be noted that the diffusion length in the low impurity concentration region is large and thus carriers can travel a long distance. Consequently, blue, green, and some portion of the red light contribute to the signal charge in this case.

This case, however, ignores the surface/interface states, which act as killers for carriers. Such states produce deep levels in the middle of the bandgap; carriers around these states are easily trapped in the levels. The lifetime in the states is generally long and trapped carriers are finally recombined there. Such trapped carriers do not contribute to the signal charge. Blue light suffers from this effect and thus, has a smaller quantum efficiency than light of longer wavelengths as shown in Fig. 2.18.

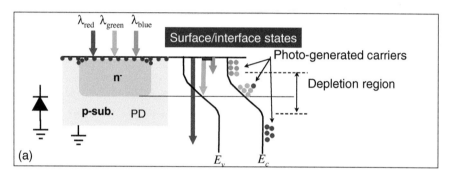

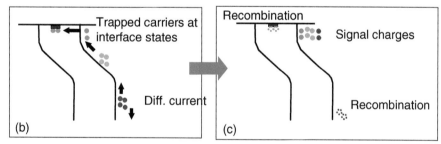

FIGURE 2.18: Behavior of photo-generated carriers in a PD with surface traps.

2.4.3.1   Pinned photodiode

To alleviate the degradation of the quantum efficiency for shorter wavelengths, the pinned photodiode (PPD) or the buried photodiode (BPD) has been developed. Historically, the PPD was first developed for CCDs [110, 111], and from the late 1990s it was adopted to CMOS image sensors [42, 112–114]. A review of PPDs can be seen in Ref. [115]. The structure of the PPD is shown in Fig. 2.19. The topmost surface of the PD has a thin p^+ layer, and thus the PD itself appears to be buried under the surface. This topmost p^+ thin layer acts to fix the Fermi level near the surface, which led to the origin of the name 'pinned photodiode'. The p^+ layer has the same potential as the p-substrate and thus the potential profile at the surface is strongly bent so that the accumulation region is separated from the surface where the trapped states are located. In this case, the Fermi level is pinned or the potential near the surface is pinned.

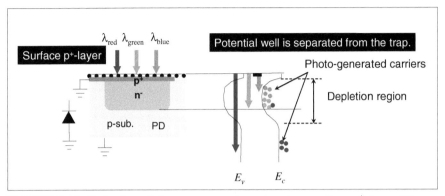

FIGURE 2.19: Behavior of photo-generated carriers in a surface p^+-layer PD or a pinned photodiode (PPD).

Eventually, the photo-generated carriers at shorter wavelengths are quickly swept to the accumulation region by the bent potential profile near the surface and contribute to the signal charge. The PPD structure has two further merits. Firstly, the PPD has less dark current than a conventional PD, because the surface p^+ layer masks the traps which are one of the main sources of dark current. Secondly, the large bent potential profile produces an accumulation region with complete depletion, which is important for 4-Tr type active pixel sensors discussed in Sec. 2.5.3. Achieving complete depletion requires not only the surface thin p^+ layer, but also an elaborate design of the potential profile by precise fabrication process control. The potential or voltage of the completely depleted region is called 'completely depleted voltage' or 'pinning voltage' [116, 117], which is an important parameter in designing 4T-APS. PPDs have been widely used for commercial CMOS image sensors with high sensitivity.

2.5   Basic pixel structures

In this section, basic pixel structures are described in detail. Historically, passive pixel sensors (PPS) were developed first, and subsequently, active pixel sensors (APS) were developed to improve the image quality. An APS has three transistors in a pixel, while a PPS has only one transistor. To achieve further improvement, an advanced APS that has four transistors in a pixel, namely the 4T-APS, has been developed. The 4T-APS has greatly improved image quality, but has a complex fabrication process.

2.5.1   Passive pixel sensor

PPS is a name coined to distinguish the above-mentioned sensors from APS, which is described in the next section. The first commercially available MOS sensor was a PPS [25, 27]; however, owing to SNR issues, its development was halted. The structure of a PPS is very simple: a pixel is composed of a PD and a switching transistor, as shown in Fig. 2.20(a). It is similar to a dynamic random access memory (DRAM).

Because of its simple structure, a PPS has a large fill factor (FF), which is the ratio of the PD area to the pixel area. A large FF is preferable for an image sensor. However, the output signal degrades easily. Switching noise is a crucial issue. In the first stage of PPS development, the accumulated signal charge was read as a current through the horizontal output line and then converted to a voltage through a resistance [25,27] or a transimpedance amplifier [28]. This scheme has the following disadvantages:

- Large smear
  Smear is a ghost signal appearing as vertical stripes without any input signal. A CCD can reduce smear. In a PPS, smear can occur when the signal charges are transferred to the column signal line. The long horizontal period (1H period, usually 64 μs) causes this smear.

- Large $k_B T C$ noise
  $k_B T C$ noise is thermal noise (discussed in detail in Sec. 2.7.1.2). Specifically, the noise power of a charge is expressed as $k_B T C$, where C is the sampling capacitance. A PPS has a large sampling capacitance of C_C in the column signal line, and hence, large noise is inevitable.

- Large column FPN
  As the capacitance of the column output line C_C is large, a column switching transistor is required for a large driving capacity, and thus, the gate size is large. This causes a large overlap gate capacitance C_{gd}, as shown in Fig. 2.20(a), which produces a large switching noise, and column FPN.

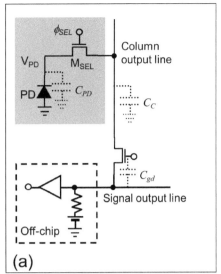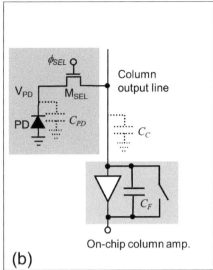

FIGURE 2.20: Basic pixel circuits of a PPS with two readout schemes. C_{PD} is a pn-junction capacitance in the PD, and C_H is a stray capacitor associated with the vertical output line. In circuit (a), an off-chip amplifier is used to convert the charge signal to a voltage signal, whereas in circuit (b), on-chip charge amplifiers or capacitive transimpedance amplifiers (CTIAs) are integrated in the column so that the signal charge can be read out almost completely. CTIAs are described in Sec. 3.2.1.3 of Chapter 3.

To address these problems, the transversal signal line (TSL) method was developed [118]. Figure 2.21 shows the concept of TSL. In the TSL structure, a column select transistor is employed in each pixel. As shown in Fig. 2.21(b), signal charges are selected in every vertical period, which are much shorter than the horizontal period. This drastically reduces smear. In addition, a column select transistor M_{CSEL} requires a small sampling capacitor C_{PD}, rather than a large capacitor C_C required for a standard PPS. Thus $k_B TC$ noise is reduced. Finally, the gate size M_{CSEL} can be reduced, and hence, little switching noise occurs in this configuration. The TSL structure has also been applied to the 3T-APS to reduce column FPN [119].

In addition, a charge amplifier on a chip with a MOS imager in place of a resistor has been reported [120]. This configuration is effective only for a small number of pixels.

Currently, a charge amplifier placed in each column is used to completely extract the signal charge and convert it into a voltage, as shown in Fig. 2.20(b). Although this configuration increases the performance, it is difficult to sense small signal charges owing to the large stray capacitance of the horizontal output line or the column output line C_C. The voltage at the column output line V_{out} is given by

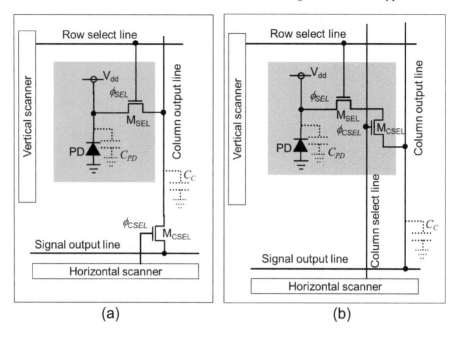

(a)   (b)

FIGURE 2.21: Improvement of PPS: (a) Conventional PPS. (b) TSL-PPS. Illustrated after [118].

$$V_{out} = Q_{PD} \frac{C_C}{C_{PD} + C_C} \frac{1}{C_F}, \qquad (2.42)$$

where Q_{PD} is the signal charge accumulated at the PD and C_{PD} is the capacitance of the PD. The stray capacitance of the horizontal output line or column output line C_C is usually much larger than C_{PD}, Eq. 2.42 is approximately expressed as,

$$V_{out} \simeq \frac{Q_{PD}}{C_F}. \qquad (2.43)$$

This implies that the charge amplifier is required to precisely convert a small charge. Present CMOS technology can integrate such charge amplifiers in each column, and thus the SNR can be improved [121]. It should be noted that this configuration consumes power.

2.5.2 Active pixel sensor, 3T-APS

This APS is named after its active element, which amplifies the signal in each pixel, as shown in Fig. 2.22. This pixel configuration is called 3T-APS, as opposed to a 4T-APS, which is described in the next section. An additional transistor M_{SF}, acts as a source follower, and thus the output voltage follows the PD voltage. The signal is transferred to a horizontal output line through a select transistor M_{SEL}.

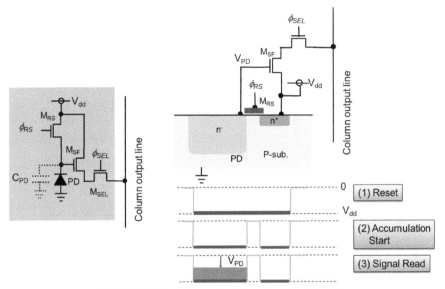

FIGURE 2.22: Basic pixel circuits of a 3T-APS.

Introducing amplification at a pixel helps the APS improve the image quality compared with the PPS. While a PPS directly transfers the accumulated signal charges to the outside of a pixel, an APS converts the accumulated signal charges to a potential in the gate. In this configuration, the voltage gain is less than one, while the charge gain is determined by the ratio of the accumulation node charge C_{PD} to a sample and the hold node charge C_{SH}.

The operation of an APS is as follows. First, the reset transistor M_{RS} is turned on. Then the PD is reset to the value $V_{dd} - V_{th}$, where V_{th} is the threshold voltage of transistor M_{RS} (see (1) Reset in Fig. 2.22). Next, M_{RS} is turned off and the PD is electrically floated ((2) Accumulation Start in Fig. 2.22). When light is incident, the photo-generated carriers accumulate in the PD junction capacitance C_{PD}. The accumulated charge changes the potential in the PD; the voltage of the PD V_{PD} decreases according to the input light intensity, as described in Sec. 2.4.1. After an accumulation time, for example, 33 ms at the video rate, the select transistor M_{SEL} is turned on and the output signal in the pixel V_{PD} is read out in the column output line ((3) Signal Read in Fig. 2.22). When the read-out process is finished, M_{SEL} is turned off and M_{RS} is again turned on to repeat the above process.

It should be noted that the accumulated signal charge is not destroyed, which makes it possible to read the signal multiple times. This is a useful characteristic for smart CMOS image sensors.

2.5.2.1   Issues with 3T-APS

Although the APS overcomes the disadvantage of the PPS, namely low SNR , there are several issues with the APS, which are as follows:

- It is difficult to suppress the $k_B TC$ noise.

- The photo-detection region, that is, the PD, simultaneously acts as a photo-conversion region. This constrains the PD design.

Here, we define the terms 'full well capacity' and 'conversion gain'. The full-well capacity is the number of charges that can be accumulated in the PD. The larger the full-well capacity, the wider the dynamic range (DR), which is defined as the ratio of the maximum output signal value V_{max} to the detectable signal value V_{min}:

$$DR = 20 \log \frac{V_{max}}{V_{min}} \text{ [dB]}. \qquad (2.44)$$

The conversion gain is defined as the voltage change when one charge (electron or hole) is accumulated in the PD. The conversion gain is thus equal to $1/C_{PD}$.

The full well capacity increases as the PD junction capacitance C_{PD} increases. On the other hand, the conversion gain, which is a measure of the increase of the PD voltage according to the amount of the accumulated charge, is inversely proportional to C_{PD}. This implies that the full-well capacity and the conversion gain have a trade-off relationship in a 3T-APS. The 4T-APS resolves the trade-off, as well as suppresses the $k_B TC$ noise.

2.5.2.2   Reset action

The usual reset operation in a 3T-APS is to turn on M_{rst} ((1) Reset in Fig. 2.22) by applying a voltage of HI or V_{dd} to the gate of M_{rst} and to fix the voltage of the PD V_{PD} at $V_{dd} - V_{th}$, where V_{th} is the threshold voltage of M_{rst}. It may be noted that in the final stage of the reset operation, V_{PD} reaches $V_{dd} - V_{th}$, and hence, the gate–source voltage across M_{rst} becomes less than V_{th}. This means that M_{rst} enters the subthreshold region. In this state, V_{PD} slowly reaches $V_{dd} - V_{th}$. This reset action is called a soft reset [122]. By employing PMOSFET with the reset transistor, this problem can be avoided, although PMOSFET consumes more area than NMOSFET because it needs an n-well area. In contrast, in a hard reset, the applied gate voltage is larger than V_{dd}, and thus M_{rst} is always above the threshold, so that the reset action finishes quickly. In this case, $k_B TC$ noise occurs as previously mentioned.

2.5.3   Active pixel sensor, 4T-APS

To alleviate the issues with the 3T-APS, the 4T-APS has been developed. In a 4T-APS, the photo-detection and photo-conversion regions are separate. Thus, the accumulated photo-generated carriers are transferred to a floating diffusion (FD) where the carriers are converted to a voltage. One transistor is added to transfer charge accumulated in the PD to the FD, making the total number of transistors in a pixel four, and hence, this pixel configuration is called 4T-APS. Figure 2.23 shows the pixel structure of the 4T-APS.

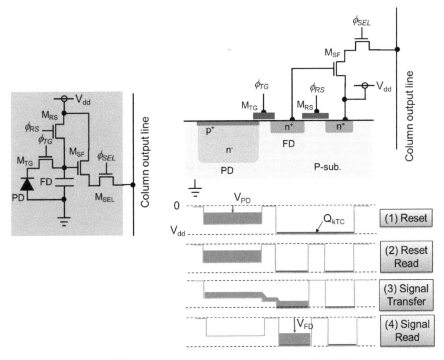

FIGURE 2.23: Basic pixel circuits of the 4T-APS.

The operation procedure is as follows. First, the signal charge accumulates in the PD. It is assumed that in the initial stage, there is no accumulated charge in the PD. Thus, a condition of complete depletion is satisfied. Just before transferring the accumulated signal charge, the FD is reset by turning on the reset transistor M_{RS}. The reset value is read out for correlated double sampling (CDS) to turn on the select transistor M_{SEL}. The CDS is described in Sec. 2.6.2.2. After the reset readout is complete, the signal charge accumulated in the PD is transferred to the FD by turning on the FD with a transfer gate M_{TG}, following the readout of the signal by turning on M_{SEL}. Repeating this process, the signal charge and reset charge are read out. It should be noted that the reset charge can be read out just after the signal charge readout. This timing is essential for CDS operation and can be realized by separating the charge accumulation region (PD) and the charge readout region (FD); this timing eliminates $k_B TC$ noise and it cannot be achieved by the 3T-APS. By this CDS operation, the 4T-APS achieves low noise operation and thus its performance is comparable to the CCDs. It should be noted that in the 4T-APS the PD must be drained of charge completely in the readout process. For this, a PPD is required. A carefully designed potential profile can achieve a complete transfer of accumulated charge to the FD through the transfer gate.

The capacitance of FD, C_{FD} determines the gain of a sensor. The smaller the C_{FD}, the higher the gain. Here, the gain means a conversion gain, g_c, that is given by $g_c = V/Q$. From this equation, $g_c = C_{FD}^{-1}$. Large conversion gain results in a small well-capacity.

2.5.3.1   Issues with 4T-APS

Although the 4T-APS is superior to the 3T-APS in its low noise level, there are some issues with the 4T-APS, as follows:

- The additional transistor reduces the FF compared to the 3T-APS.
- Image lag may occur when the accumulated signal charge is completely transferred into the FD.
- It is difficult to establish fabrication process parameters for the PPD, transfer gate, FD, reset transistor, and other units, for low noise and low image lag performance.

Figure 2.24 illustrates incomplete charge transfer in a 4T-APS. In Fig. 2.24(a), the charges are completely transferred to the FD, while in Fig. 2.24(b) some charge remains in the PD, causing image lag. To prevent incomplete transfer, elaborate potential profiles are required [123, 124].

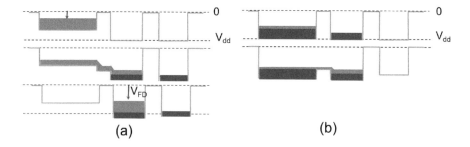

(a)   (b)

FIGURE 2.24: Incomplete charge transfer in a 4T-APS. (a) Complete transfer. PD is depleted completely. (b) Incomplete transfer. Causes of image lag and noise.

2.5.3.2   Pixel sharing

Although 4T-APS is effective for higher SNR than 3T-APS, larger pixel area is occupied due to the addition of one transistor, that is, a transfer transistor, in a pixel. This is a critical disadvantage in a pixel pitch compared with the CCD where only one transfer gate is required in a pixel described in Sec. 2.10.

To alleviate this, pixel sharing techniques have been developed. Some parts in a pixel, for example FD, can be shared with each other so that the pixel size can be reduced [128]. Figure 2.25 shows some examples of pixel sharing schemes.

The FD driving sharing technique [127] shown in Fig. 2.25(d) is used to reduce the number of transistors in a 4T-APS by one [129]. The select transistor can be

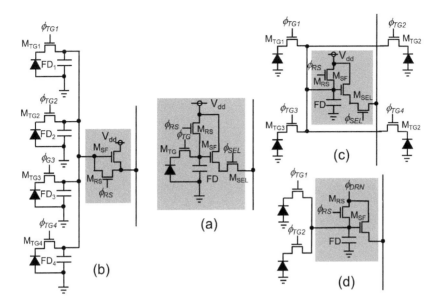

FIGURE 2.25: Pixel sharing. (a) Conventional 4T-APS. (b) Sharing of a select transistor and a source follower transistor. Illustrated after [125]. (c) Pixels with only a PD and transfer gate transistor while the other elements including the FD are shared. Illustrated after [126]. (d) As in (c) but with the reset voltage controlled. Illustrated after [127].

eliminated by controlling the potential of the FD by changing the pixel drain voltage through the reset transistor. Recently, sensors have been reported with around 1-μm pitch pixels using pixel sharing technology [130, 131].

2.6 Sensor peripherals

2.6.1 Addressing

In CMOS image sensors, to address each pixel, a scanner or a decoder is used. A scanner consists of a latch array or shift register array to carry data in accordance with a clock signal. When using scanners with vertical and horizontal access, the pixels are sequentially addressed. To access an arbitrary pixel, a decoder, which is a combination of logic gates, is required. A decoder arbitrarily converts N input data to 2^N output data using customized random logic circuits. Figure 2.26 shows a typical scanner and a decoder. Figure 2.27 presents an example of a decoder, which decodes 3-bit input data to six output data.

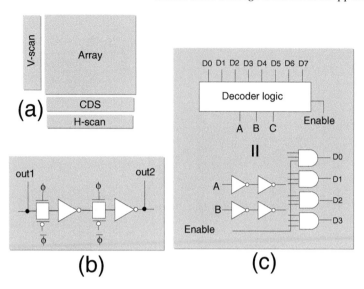

FIGURE 2.26: Addressing methods for CMOS image sensors: (a) sensor architecture, (b) scanner, (c) decoder.

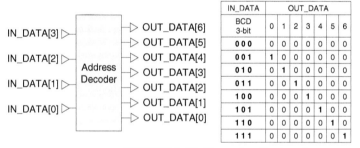

FIGURE 2.27: Example of a decoder.

Random access An advantage of smart CMOS image sensors is the random access capability, wherein an arbitrary pixel can be addressed at any time. A typical method to implement random access is to add one transistor to each pixel so that a pixel can be controlled with a column switch, as shown in Fig. 2.28(a). Row and column address decoders are also required instead of scanners, as mentioned above. It should be noted that if an extra transistor is added in series with the reset transistor, as shown in Fig. 2.28(b), then anomalies will occur for some duration [132]. In this case, if M_{RRS} is turned on, the accumulated charge in the PD is distributed between the PD capacitance C_{PD} and a parasitic capacitance C_{diff}, which degrades the signal charge as shown in Fig. 2.28 (c).

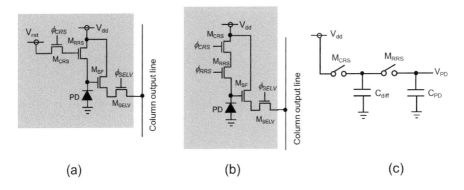

(a) (b) (c)

FIGURE 2.28: Two types of pixel structures for random access, (a) and (b). (c) Equivalent circuits of (b).

Multi-resolution Multi-resolution is another addressing technique for CMOS image sensors [70,133]. It is a method to vary the resolution of a sensor; for example, in a video graphics array (VGA, 640 × 480-pixel) sensor, the resolution can be changed by a factor of 1/4 (320 × 240-pixel), a factor of 1/8 (160 × 120-pixel), etc. To quickly locate an object with a sensor, a coarse resolution is effective as the post-image processing load is low. This is effective for target tracking, robotics, etc.

2.6.2 Readout circuits

2.6.2.1 Source follower

The voltage of a PD is read with a source follower (SF). As shown in Fig. 2.29, a source follower transistor M_{SF} is placed in a pixel and a current load M_b is placed in each column. A select transistor M_{SEL} is located between the follower and the load. It should be noted that the voltage gain A_v of an SF is less than 1 and is expressed by the following:

$$A_v = \frac{1}{1 + g_{mb}/g_m},$$ (2.45)

where g_m and g_{mb} are the transconductance and body transconductance of M_{SF}, respectively [134]. The DC response of an SF is not linear over the input range. The output voltage is sampled and held in the capacitance for CDS C_{CDS}. The CDS is mentioned in the next section, Sec. 2.6.2.2.

   In the readout cycle using an SF, the charge and discharge processes associated with a S/H capacitor C_{SH} are the same. In the charge process, C_{SH} is charged with a constant voltage mode so that the rise time t_r is determined by the constant voltage mode. In the discharge process, C_H is discharged with a constant current mode by the current source of the SF, and hence, the fall time t_f is determined by the constant current mode. The inset of Fig. 2.29 illustrates this situation. These characteristics must be evaluated when the readout speed is important [135].

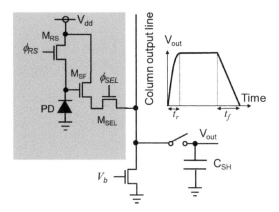

FIGURE 2.29: Readout circuits using a source follower. The inset shows the dependence of output voltage V_{out} dependence on time in the readout cycle. Illustrated after [135].

2.6.2.2 Correlated double sampling

Correlated double sampling (CDS) is used to eliminate the thermal noise generated in a reset transistor of the PD, which is the $k_B TC$ noise. Several types of CDS circuitry have been reported and are reviewed in detail in Ref. [3]. Table 2.2 summarizes the CDS types following the classification of Ref. [3].

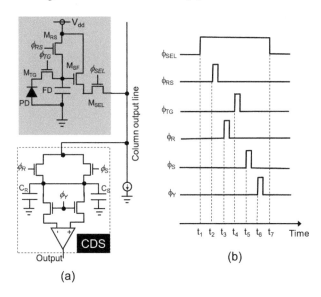

FIGURE 2.30: CDS: (a) basic circuits and (b) timing chart.

Figure 2.30 shows a typical circuitry for CDS with an accompanying 4T-APS type pixel circuit. The basic CDS circuit consists of two sets of S/H circuits and a differential amplifier. The reset and signal levels are sampled and held in the capacitances C_R and C_S, respectively, and then the output signal is produced by differentiating the reset and signal values held in the two capacitors.

The operating principle can be explained as follows with the help of the timing chart in Fig. 2.30(b). In the signal readout phase, the select transistor M_{SEL} is turned on from t_1 to t_7 when Φ_{SEL} turns on ("HI"–level). The first step is to read the reset level or $k_B TC$ noise and store it in the capacitor C_R just after the FD is reset at t_2 by setting Φ_{RS} to HI. To sample and hold the reset signal in the capacitor C_R, Φ_R becomes HI at t_3. The next step is to read the signal level. After transferring the accumulated signal charge to the FD by turning on the transfer gate of M_{TG} at t_4, the accumulated signal is sampled and held in C_S by setting Φ_S to HI. Finally, the accumulated signal and the reset signal are differentiated by setting Φ_Y to HI.

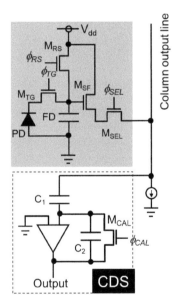

FIGURE 2.31: An alternative circuit for CDS. Here a capacitor is used to subtract the reset signal.

Another CDS circuit is shown in Fig. 2.31 [136, 137]. Here, the capacitor C_1 is used to subtract the reset signal as a clamp capacitor. M_{CAL} acts as a switch for auto-zero operation of the OP amplifier. When the transistor M_{CAL} turns ON, the offset voltage V_{OS} of the operational amplifier (OPA) is sampled in the capacitor C_1. In that case, the reset noise V_{RST} is also applied to C_1. Thus, the applied voltage of C_1, i.e., ΔV is given by $= V_{RST} + V_{OS}$. After M_{CAL} turns OFF, the transfer gate M_{TX} is turned on and the pixel signal flows into the FD. Consequently, the charge in the FD is the sum of the signal charge and reset noise charge mentioned in Sec. 2.5.3.

In that case, $V_{SIG} + V_{RST}$ is applied to the one node of C_1, and the other node of C_1, that is, the input node of the OPA, changes to $V_{SIG} - V_{OS}$, because the accumulated charge, ΔV does not change. This voltage, $V_{SIG} - V_{OS}$ implies that the reset noise is subtracted as CDS. The input offset of the OPA is canceled, that is, the auto-zero operation is achieved.

TABLE 2.2
CDS types for CMOS image sensors

Category	Method	Feature	Ref.
Column CDS 1	One coupling capacitor	Simple structure but suffers from column FPN	[136]
Column CDS 2	Two S/H capacitors	ibid.	[138]
DDS*	DDS following column CDS	Suppression of column FPN	[139]
Chip-level CDS	I–V conv. and CDS in a chip	Suppression of column FPN but needs fast operation	[114]
Column ADC	Single slope ADC	Suppression of column FPN	[140, 141]
	Cyclic ADC	ibid.	[142]

*double delta sampling.

2.6.3   Analog-to-digital converters

In this section, analog-to-digital converters (ADCs) for CMOS image sensors are briefly described. For sensors with a small number of pixels, such as QVGA (230×320) and CIF (352×288), a chip-level ADC is used [143, 144]. When the number of pixels increases, column parallel ADCs are employed, such as a successive approximation register (SAR) ADC [145, 146], a single slope (SS) ADC [147, 148], a cyclic (CY) ADC [149, 150], and $\Delta\Sigma$ ADC [151, 152]. A review of column parallel ADCs in CMOS image sensors is available in [153].

SAR ADCs can achieve high speed conversion with relatively low power consumption and simple structure. They are required in digital-to-analog converters (DACs), which are area-consuming. SS ADCs are conventionally used for column parallel ADCs because they are efficient in space utilization. The resolution is low to medium, and the speed is rather low because 2^N clocks are required to obtain the N-bit. Cyclic ADCs can achieve high resolution and high speed; however, they are both area- and power-consuming. The disadvantage of high area-requirement can be alleviated by introducing 3D stacked structure [154], which is described in Sec. 3.3.4 of Chapter 3. A multiple-sampling is compatible to the architecture of a cyclic ADC and thus, is often combined with the ADC to achieve high sensitivity. The features of delta-sigma ADCs are similar to those of cyclic ADCs. Fine process technology can allow these ADCs to be used for column-parallel architecture.

The comparison among the four ADCs is summarized in Table 2.3. Furthermore, there have been numerous reports on pixel-level ADCs [63, 155, 156]. ADCs are employed in applications that have the same point of view as the architecture of smart CMOS image sensors, namely, as pixel-level, column-level, and chip-level. Among these ADCs, SAR and SS ADCs are widely used for column ADCs for CMOS image sensors. In the following section, SAR and SS ADCs are introduced.

TABLE 2.3
ADCs for CMOS image sensors

Method	SAR	SS	Cyclic	$\Delta\Sigma$
Area Efficiency	Low	High	Low	Medium
Conversion Speed	High ($N \times T_{clk}$)	Low ($2^N \times T_{clk}$)	High ($N \times T_{clk}$)	Medium
Resolution	High	Medium	High	High
Energy Efficiency	Medium	Low ~Medium	Medium	High

Successive approximation register (SAR) ADC   The successive approximation register (SAR) ADC is composed of a comparator, SAR logic, and N-bit digital-to-analog converter (DAC) as shown in Fig. 2.32.

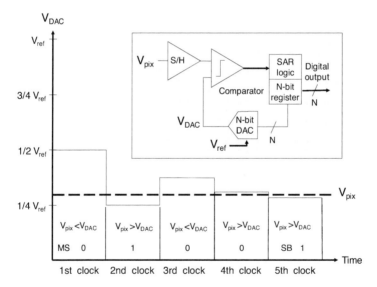

FIGURE 2.32: The basic block of a successive approximation register (SAR) ADC. The timing chart is inserted to explain the sequence of successive approximation logic.

In this figure, a 5-bit ADC operation is shown. The pixel analog signal V_{pix} is input to the sample and holds (S/H) circuits. The N-bit register is set to MSB=1 and the others to 0, which corresponds to $V_{DAC} = V_{ref}/2$. V_{ref} is the reference voltage of the DAC. V_{pix} is compared with this value, V_{DAC}, and if $V_{pix} > V_{DAC}$, MSB=1; otherwise MSB=0. In the next clock, the next bit is set to 1, and V_{pix} is compared with V_{DAC}. In

Fig. 2.32, $V_{pix} < V_{DAC}$, then the next bit =0. Repeating this process, all the bits are determined as shown in Fig. 2.32.

The conversion speed of N-bit SAR ADC is $N \times T_{clk}$, which is high compared to other ADCs. A disadvantage of SAR ADC for CMOS image sensors is its low area efficiency owing to the DAC circuits.

Single slope (SS) ADC   These have been widely used in column ADCs for CMOS image sensors, because their high area efficiency.   The operating principle of SS-ADC is simple. It consists of a comparator, counter, and ramp signal generator (RSG) as shown in Fig. 2.33. RSG is commonly used for all column ADCs. The input analog signal V_{pix} is compared with V_{ramp}, which is the ramp signal from RSG. When V_{ramp} becomes larger than V_{pix}, the comparator and the counter output digital values. The digital CDS based on SS-ADC is described in Sec. 4.2.5 of Chapter 4.

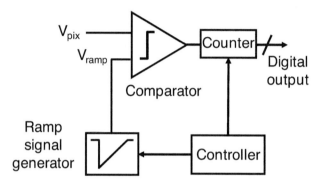

FIGURE 2.33: The basic block of single slope (SS) ADC.

2.7   Basic sensor characteristics

In this section, some basic sensor characteristics are described.   For details on measurement techniques of image sensors, please refer to Refs. [157, 158].

2.7.1   Noise

2.7.1.1   Fixed pattern noise

In an image sensor, spatially fixed variations of the output signal are of great concern for image quality.   This type of noise is called the fixed pattern noise (FPN). Regular variations, such as column FPN, can be perceived more easily than random variations.   A variation of 0.5% of pixel FPN is considered acceptable threshold values, while 0.1% of column FPN is acceptable [159]. Employing column amplifiers sometimes causes column FPN. In Ref. [159], the column FPN is suppressed by randomizing the relation between the column output line and the column amplifier.

2.7.1.2 Temporal noise

kTC noise   In a CMOS image sensor, the reset operation mainly causes thermal noise. When the accumulated charge is reset through a reset transistor, the thermal noise $4k_BTR_{on}\delta f$ is sampled in the accumulation node, where δf is the frequency bandwidth and R_{on} is the ON-resistance of the reset transistor, as shown in Fig. 2.34. The accumulation node is a PD junction capacitance in a 3T-APS and an FD capacitance in a 4T-APS.

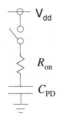

FIGURE 2.34: Equivalent circuits of kTC noise. R_{on} is the ON-resistance of the reset transistor and C_{PD} is the accumulation capacitance, which is a PD junction capacitance for a 3T-APS and a floating diffusion capacitance for a 4T-APS, respectively.

The thermal noise is calculated to be k_BT/C_{PD}, which does not depend on the ON-resistance R_{on} of the reset transistor. This is because larger values of R_{on} increase the thermal noise voltage per unit bandwidth, while simultaneously decreasing the bandwidth [160], which masks the dependence of R_{on} on the thermal noise voltage. We now derive a formula for the thermal noise voltage referring to the configuration in Fig. 2.34. The thermal noise voltage is expressed as

$$\overline{v_n^2} = 4k_BTR_{on}\Delta f. \tag{2.46}$$

As shown in Fig. 2.34, the transfer function is expressed as

$$\frac{v_{out}}{v_n}(s) = \frac{1}{R_{on}C_{PD}s+1}, \quad s = j\omega. \tag{2.47}$$

Thus, the noise is calculated as

$$\begin{aligned} \overline{v_{out}^2} &= \int_0^\infty \frac{4k_BTR_{on}}{(2\pi R_{on}Cf)^2+1}df \\ &= \frac{k_BT}{C}. \end{aligned} \tag{2.48}$$

The noise power of the charge q_{out}^2 is expressed as

$$q_{out}^2 = (Cv_{out})^2 = k_B T C. \tag{2.49}$$

The term "kTC" noise originates from this formula. The $k_B T C$ noise can be eliminated by the CDS technique. However, it can be applied only to a 4T-APS, and is difficult to apply to a 3T-APS.

1/f noise   1/f noise is so named because the noise level is inversely proportional to the frequency. It is sometimes called flicker noise. In MOSFETs, defects in channel interface produce traps, which cause a carrier number to temporarily fluctuate, and the fluctuation can be observed as a noise. The noise in MOSFET is empirically expressed as [161],

$$v_n^2 = \frac{K_F}{C_{ox}^2 WL} \frac{1}{f}, \tag{2.50}$$

where K_F is equal to 5×10^{-9} fC$^2/\mu$m^2 for NMOSFETs and 2×10^{-10} fC$^2/\mu$m^2 for PMOSFETs [162]. Thus 1/f noise in PMOSFETs is smaller than that in NMOSFETs. As shown in Eq. 2.50, the noise is inversely proportional to the total gate area of MOSFET. Therefore, the 1/f noise is a cause for concern primarily in small size transistors, i.e., in a small-sized pixel.

RTS noise   Random telegraph signal (RTS) noise originates from the random capture and emission of electrons in the interface states of the gate oxide of MOS transistors. It is also called popcorn noise or burst noise, as the noise is observed as discrete drain current fluctuations in small size MOSFETs. It is reported that 1/f noise is produced by superimposed RTSs with various time constants. In an RTS, the noise is inversely proportional to the square of the frequency over specific frequency.

Shot noise   As described in Sec. 2.3.1.4, shot noise is proportional to the square root of the number of photons. Thus, the signal to noise ratio (SNR) is proportional to the square root of the number of photons. Figure 2.35 shows the dependence of the number of electrons on signal and noise on exposure amount (illuminance).

2.7.2   Dynamic range

The dynamic range (DR) of an image sensor is defined as the ratio of the output signal range to the input signal range. DR is thus determined by two factors, the noise floor and the well charge capacity called 'full well capacity'. In Fig. 2.35, DR is defined between the noise floor and the saturation level. The saturation level is determined by the full well capacity. It should be noted that optical DR and the output DR are defined as shown in Fig. 2.35. Most of the sensors have almost the same DR of around 70 dB, which is mainly determined by the well capacity of the PD. For some applications, such as in automobiles, this value of 70 dB is not sufficient and a DR of over 100 dB is required. Considerable efforts have been made to enhance the DR and these are described in Chapter 4.

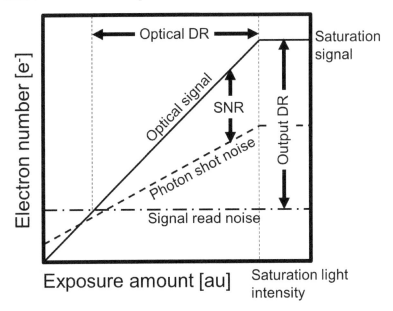

FIGURE 2.35: Electron number as a function of exposure amount. In the low light level, the signal read noise including the dark noise is dominant, while in the high light level, the photon shot noise is dominant.

2.7.3 Speed

One of the causes to limit the speed of an APS is the diffusion carriers. Some of the photo-generated carriers in the deep region of a substrate will finally arrive at the depletion region, acting as slow output signals. The diffusion time for electrons and holes as a function of impurity concentration is shown in Fig. 2.4. It should be noted that the diffusion lengths for both holes and electrons are over a few tens of μm and sometimes reach a few hundreds of μm. Thus, a careful treatment is needed to achieve high speed imaging. This effect greatly degrades the PD response, especially in the near infrared (NIR) region. To alleviate this effect, some structures prevent diffusion carriers from entering the PD region. To enhance the response speed of PD, it is effective to introduce drift mechanism inside the PD.

CR time constants are another major factor limiting the speed, because the vertical output line in smart CMOS image sensors is generally so long that the associated resistance and stray capacitance are large. The requirements for the high speed operation of CMOS image sensors are discussed in Sec. 4.3 of Chapter 4.

2.8   Color

There are three ways to realize color in a conventional CMOS image sensor, as shown in Fig. 2.36. They are explained in the following sections.

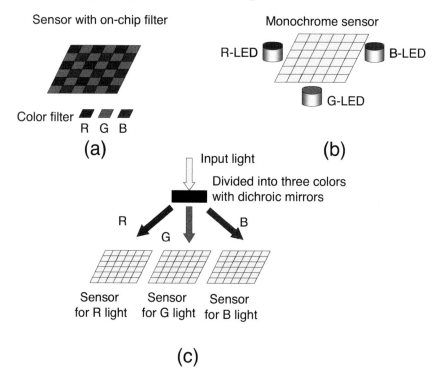

FIGURE 2.36: Methods to realize color in CMOS image sensors: (a) On-chip color filter, (b) three light sources, and (c) three image sensors.

Although color is an important characteristic for general CMOS image sensors, the implementation method is almost the same as that for smart CMOS image sensors and a detailed discussion is beyond the scope of this book. Color treatment for general CMOS image sensors is described in detail in Ref. [2]. Sec. 3.3.5 of Chapter 3 details selected topics on realizing colors using smart functions. The wavelength region of visible region is summarized in Appendix D.1.

2.8.1   On-chip color filter type

Three colored filters are directly placed on the pixels. Typically red (R), green (G), and blue (B) (RGB) or CMY complementary color filters of cyan (Cy), magenta

(Mg), and yellow (Ye) are used. The representation of CMY and RGB is as follows (W indicates white):

$$Ye = W - B = R + G,$$
$$Mg = W - G = R + B, \quad (2.51)$$
$$Cy = W - R = G + B.$$

Usually, the color filters are made of an organic film, which is composed of pigment; however, inorganic color film has also been used to make these filters [163]. The thickness of α-Si is controlled to produce a color response. This helps reduce the thickness of the color filters, which is important in optical crosstalk in a fine pitch pixel less than 2 μm in pitch.

Bayer pattern The Bayer pattern is commonly used to place the three RGB filters [164]. This type of on-chip filter is widely used in commercially available CMOS image sensors. To retrieve an original image, demosaicking process is required [165].

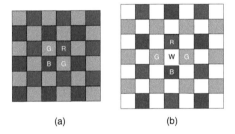

(a)   (b)

FIGURE 2.37: On-chip filter pattern: (a) Bayer pattern, and (b) RGB plus white (no filter) pattern.

White pixel When the pixel size is shrunk, the total input light intensity is reduced and thus, SNR may be degraded. Introducing white pixels is effective in improving the SNR. A white pixel means no color filter and hence, almost no absorption exists in white pixel. To obtain R, G, and B colors in RGBW pixels, numerous conversion methods are proposed. One example is shown in Fig. 2.37 (b). From Ref. [166], the conversion equations are as follows:

$$R_W = W \frac{R_{av}}{R_{av} + G_{av} + B_{av}},$$
$$G_W = W \frac{G_{av}}{R_{av} + G_{av} + B_{av}},$$
$$B_W = W \frac{B_{av}}{R_{av} + G_{av} + B_{av}}$$

Inorganic color filter   Conventional color filters are made of organic materials. To improve their durability, inorganic color filters have been developed [167]. In Fig. 2.38, a photonic crystal or an interference film made of SiO_2/TiO_2 multi-layer is fabricated on the surface of the pixel array. In this case, a near infrared (NIR) filter is also fabricated and hence, this sensor can be used not only in daytime, but also at night.

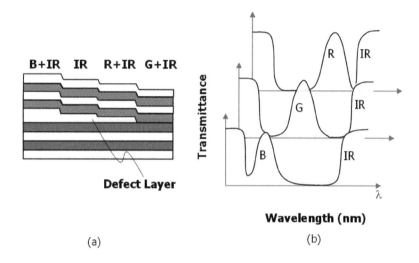

(a)   (b)

FIGURE 2.38: Color filter based on photonic crystal. Color with NIR detection can be achieved. Adapted from [167] with permission.

2.8.2   Three imagers type

In the three imagers method, three CMOS image sensors without color filters are used for R, G, and B. To divide the input light into three colors, two dichroic mirrors are used. This configuration realizes high-color fidelity; however, it requires complicated optics and is expensive. It is usually used in broadcasting systems, which require high-quality images.

2.8.3   Three light sources type

This uses artificial RGB light sources, with each RGB source illuminating the objects sequentially. One sensor acquires three images for the three colors, with the three images being combined to form the final image. This method is mainly used in medical endoscopes. The color fidelity is excellent; however, the time to acquire a whole image is longer than that for the above two methods. This type of color representation is not applicable to conventional CMOS image sensors because they usually have a rolling shutter. This is discussed in Sec. 5.5 of Chapter 5.

2.9 Comparison among pixel architectures

In this section, several types of pixel architectures, namely PPS, 3T-APS, and 4T-APS, as well as the log sensor which is detailed in Chapter 3 are summarized in Table 2.4. At present, the 4T-APS has the best performance with regard to noise characteristics and is widely used in CMOS image sensors. However, it should be noted that other systems have their own advantages, which provide possibilities for smart sensor functions.

TABLE 2.4
Comparison among PPS, 3T-APS, 4T-APS, and the log sensor. The log sensor is discussed in Chapter 3.

	PPS	3T-APS	4T-APS (PD)	4T-APS (PG)	Log
Sensitivity	Depends on the performance of a charge amp	Good	Good	Fairly good	Good but poor at low light level
Area consumption	Excellent	Good	Fairly good	Fairly good	Poor
Noise	Fairly good	Fairly good (no kTC reduction)	Excellent	Excellent	Poor
Dark current	Good	Good	Excellent	Good	Fairly good
Image lag	Fairly good	Good	Fairly good	Fairly good	Poor
Process	Standard	Standard	Special	Special	Standard
Note	Very few commercialized	Widely commercialized	Widely commercialized	Very few commercialized	Recently commercialized

2.10 Comparison with CCDs

In this section, CMOS image sensors are compared with CCDs. The fabrication process technologies of CCD image sensors have been developed only for the CCD image sensors themselves, while those of CMOS image sensors were originally developed for standard mixed signal processes. Although the recent development of CMOS image sensors requires dedicated fabrication process technologies, CMOS image sensors are still based on standard mixed signal processes. The stacked technology described in Sec. 3.3.4 may alleviate this restriction and make it possible to select a process technology suitable for each of the image sensors and the signal processing circuits.

There are two main differences between the architecture of CCD and CMOS sensors, the signal transferring method and the signal readout method. Figure 2.39 illustrates the structures of CCD and CMOS image sensors.

TABLE 2.5
Comparison between CCD and a CMOS image sensors

Item	CCD image sensor	CMOS image sensor
Readout scheme	One on-chip SF; limits speed	SF in every column; may exhibit column FPN
Simultaneity	Simultaneous readout of every pixel	Sequential reset for every row; rolling shutter
Transistor isolation	Reverse biased pn-junction	LOCOS/STI[†]; may exhibit stress-induced dark currents
Thickness of gate oxide	Thick for complete charge transfer ($>$ 50 nm)	Thin for high speed transistor and low voltage power supply ($<$ 10 nm)
Gate electrode	Overlapped 1st & 2nd poly-Si layers	Polycide poly-Si
Isolation layers	Thin for suppressing light guide	Thick ($\sim 1\mu$m)
Metal layer	Usually one	Over three layers
# Transistors	1 (Transfer gate)	4 (4T-APS)

[†]LOCOS: local oxidation of silicon, STI: shallow trench isolation.

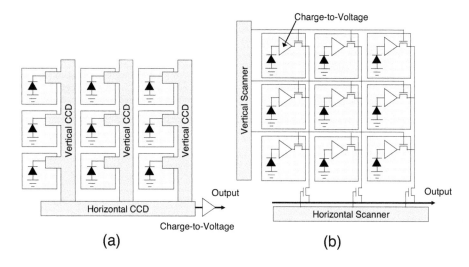

FIGURE 2.39: Conceptual illustration of the chip structure of (a) CCD image sensor and (b) CMOS image sensor.

A CCD transfers the signal charges to the end of the output signal line as received and converts them into a voltage signal through an amplifier. In contrast, a CMOS image sensor converts the signal charge into a voltage signal at each pixel. The in-pixel amplification may cause FPN and thus the quality of early CMOS image sensors was worse than that of CCDs. However, this drawback has been drastically improved. In a high-speed operation, the in-pixel amplification configuration gives better gain-bandwidth than a configuration with one amplifier on a chip.

In CCD image sensors, the signal charge is transferred simultaneously, which gives low noise and high power consumption. Furthermore, this signal transfer gives the same accumulation time for every pixel at any time. In contrast, in CMOS image sensors, the signal charge is converted at each pixel and the resultant signal is read out row-by-row, and as a result, the accumulation time is different for pixels in different rows at any time. This is referred to as a "rolling shutter."

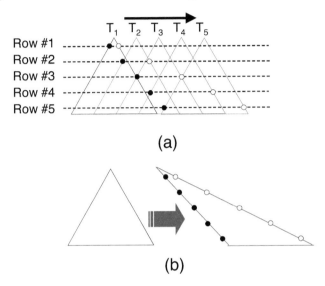

(a)

(b)

FIGURE 2.40: Illustration of a rolling shutter. (a) A triangle-shape object moves from left to right. (b) The original image is distorted.

Figure 2.40 illustrates the origin of the rolling shutter. A triangle shape object moves from the left to the right. In the imaging plane, the object is scanned row-by-row. In Fig. 2.40(a) at $Time_k$ (k = 1, 2, 3, 4, and 5), the sampling points are shown in Rows #1–#5. The original figure (left in Fig. 2.40(b)) is distorted in the detected image (right in Fig. 2.40(b)), which is constructed from the corresponding points in Fig. 2.40(a).

Recently, the global shutter architecture has been introduced in 4T-APS CMOS image sensors, which is commercialized. Although the global shutter is still not common in conventional CMOS image sensors, it is expected to be widely introduced in many CMOS image sensors. The details are described in Sec. 4.3.2. Table 2.5 summarizes the comparison between the CCD and CMOS image sensors, including these features.

When the pixel pitch in a smartphone camera is required to be reduced, the CMOS image sensors have two types of disadvantages compared to the CCDs. Firstly, while there are four transistors in the 4T-APS, there is only one transfer gate in the CCD as shown in Figs. 2.41(a) and (b). To alleviate this issue, pixel sharing techniques have been developed as described in Sec. 2.5.3.2. Secondly, the optical crosstalk becomes larger in CMOS image sensors when the pixel pitch is shrunk. Because the distance between the PD surface and the micro-lens in CMOS image sensors is larger than that in the CCDs, several interconnection layers exist. On the contrary, the CCDs have no such interconnection layers and thus, only a small gap exists between the PD surface and the micro-lens as shown in Figs. 2.41(c) and (d). A backside illumination (BSI) structure, which is described in Sec. 3.3.3, can realize almost the same thickness as does a CCD. Hence, the optical crosstalk is drastically reduced even in a small pixel pitch such as 1 μm pitch.

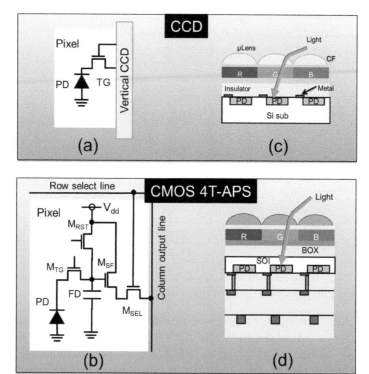

FIGURE 2.41: Comparison between CCD and CMOS in terms of pixel structure (a) and (b), and in terms of optical crosstalk (c) and (d), respectively.

3

Smart structures and materials

3.1 Introduction

Smart CMOS image sensors can employ smart pixels, structures, and materials on their chips. In addition, dedicated pixel arrangement and optics are incorporated in a smart CMOS image sensor. In this chapter, an overview of various types of smart pixels, structures, and materials is presented. Next, dedicated pixel arrangement and optics for smart CMOS image sensors are overviewed.

In a conventional CMOS image sensor, the output from a pixel is an analog voltage signal with a source follower (SF), giving an analog voltage output, as mentioned in Sec. 2.6.2.1. To realize some of the smart functions, however, other kinds of modes such as the analog current mode, pulse output mode, and digital output mode, have been developed. In the subsequent sections, these three types of output modes are introduced as smart pixels, starting with the analog output mode. The pulse and digital output modes are subsequently presented. The pulse output mode is a mixture of analog and digital processing.

In the second part of this chapter, structures and materials other than standard silicon CMOS technologies are introduced for certain CMOS image sensors. The recent advances in LSI technologies have introduced many new materials and structures such as silicon-on-insulator (SOI), silicon-on-sapphire (SOS), and 3D integration, and the use of many other materials, such as silicon-germanium (SiGe) and Ge. These new structures and materials in a smart CMOS image sensor can enhance its performance and functions as enumerated in Table 3.1.

TABLE 3.1
Structures and materials for smart CMOS image sensors

Structure/material	Features
SOI	Small area when using both NMOS and PMOS
SOS	Transparent substrate (sapphire)
3D integration	Large FF, integration of signal processing
Multi-path, SiGe/Ge	Long wavelength (NIR)

In the last part of this chapter, dedicated pixel arrangement and optics for smart CMOS sensors are mentioned.

3.2 Smart pixels

In this section, various pixel structures for smart CMOS image sensors are introduced as smart pixels. These structures are different from the conventional active pixel sensor (APS) structure. The smart pixels are classified into three types from the viewpoint of output modes. Firstly, the analog mode is introduced, followed by pulse modulation mode, which includes both pulse width modulation and pulse frequency modulation. Finally, SPAD, which is introduced in Sec. 2.3.4.1, is explained in detail.

3.2.1 Analog mode

APS is a typical pixel in analog processing. In the following subsections, in addition to the conventional APS types, current mode APS, Log sensor, CTIA (Capacitive Transimpedance Amplifier) pixel, and lock-in pixel are introduced.

3.2.1.1 Current mode

The conventional APS outputs a signal as a voltage. For signal processing, the current mode is more convenient, because signals can easily be summed and subtracted by Kirchhoff's current law. For an arithmetic unit, multiplication can be easily employed with a current mirror circuit, which is also used to multiply the photocurrent through a mirror ratio greater than one. It is noted however that this causes pixel fixed pattern noise (FPN).

In the current mode, the memory can be implemented using a current copier circuit [168]. FPN suppression and analog-to-digital converters (ADCs) in current mode have also been demonstrated [169]. The current mode is classified into two categories: direct output mode and accumulation mode.

Direct mode   In the direct output mode, the photocurrent is directly output from the photodetector, which could be a photodiode (PD) or a photo-transistor (PTr) [170, 171]. The photocurrent from a PD is usually transferred by current mirror circuits with or without current multiplication. Some early smart image sensors used current mode output with a photo-transistor and a current mirror. The mirror ratio is used to amplify the input photocurrent, as described above. The architecture however suffers from low sensitivity at low light levels and large FPN owing to a mismatch in the current mirror. The basic circuit is shown in Fig. 3.1.

Accumulation mode   Figure 3.2 shows the basic pixel structure of a current mode APS [172,173]. By introducing the APS configuration, the image quality is improved compared with the direct mode. The output of the pixel is expressed as follows:

$$I_{pix} = g_m \left(V_{gs} - V_{th}\right)^2, \tag{3.1}$$

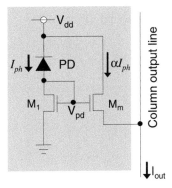

FIGURE 3.1: Basic circuit of a pixel using a current mirror. The ratio of the transistor W/L in M_1 to M_m is α, so that the mirror current or the output current is equal to αI_{ph}.

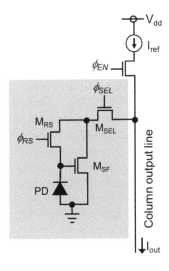

FIGURE 3.2: Basic circuit of a current mode APS. Illustrated after [172].

where V_{gs} and g_m are the gate-source voltage and the transconductance of the transistor M_{SF}, respectively. At the reset, the voltage at the PD node is

$$V_{reset} = \sqrt{\frac{2L_g}{\mu_n C_{ox} W_g} I_{ref}} + V_{th}. \qquad (3.2)$$

As light is incident on the PD, the voltage at the node becomes

$$V_{PD} = V_{reset} - \Delta V, \qquad (3.3)$$

where T_{int} is the accumulation time and ΔV is

$$\Delta V = \frac{I_{ph}T_{int}}{C_{PD}},\tag{3.4}$$

which is the same as that for a voltage mode APS. Consequently, the output current is expressed as

$$I_{pix} = \frac{1}{2}\mu_n C_{ox}\frac{W_g}{L_g}\left(V_{reset} - \Delta V - V_{th}\right)^2.\tag{3.5}$$

The difference current $I_{diff} = I_{ref} - I_{pix}$ is then

$$I_{diff} = \sqrt{2\mu_n C_{ox}\frac{W_g}{L_g}I_{ref}\Delta V - \frac{1}{2}\mu_n C_{ox}\frac{W_g}{L_g}\Delta V^2}.\tag{3.6}$$

It is noted that the threshold voltage of the transistor M_{SF} is canceled so that the FPN originating from the variation of the threshold could be improved. Further discussion on this topic appears in Ref. [172].

3.2.1.2 Log sensor

A conventional image sensor responds linearly to the input light intensity. A log sensor is based on the sub-threshold operation mode of MOSFET. Appendix F explains the sub-threshold operation. A log sensor pixel uses the direct current mode, because the current mirror configuration has a log sensor structure when the photocurrent is so small that the transistor enters the sub-threshold region. Log sensors are mainly applied to wide dynamic range image sensors [174–177]. It was initially proposed and demonstrated in 1984 by Chamberlain and Lee [178]. Wide dynamic range image sensors are described in Sec. 4.4.

Figure 3.3 shows the basic pixel circuit of a logarithmic CMOS image sensor. In the sub-threshold region, the MOSFET drain current I_d is very small and exponentially increases with the gate voltage V_g:

$$I_d = I_o \exp\left(\frac{e}{mk_BT}\left(V_g - V_{th}\right)\right).\tag{3.7}$$

For the derivation of this equation and the meaning of the parameters, refer to Appendix F.

In the log sensor of Fig. 3.3(b),

$$V_G = \frac{mkT}{e}\ln\left(\frac{I_{ph}}{I_o}\right) + V_{ps} + V_{th}.\tag{3.8}$$

In this sensor the accumulation mode is employed in the log sensor architecture. For a drain current of M_c, I_c is expressed as

$$I_c = I_o \exp\left[\frac{e}{mk_BT}\left(V_G - V_{out} - V_{th}\right)\right].\tag{3.9}$$

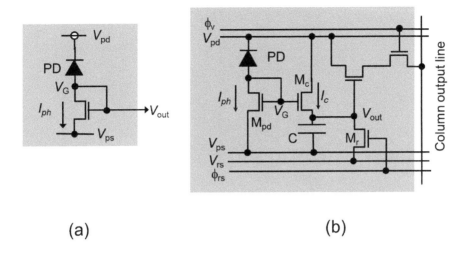

(a)   (b)

FIGURE 3.3: Pixel circuit of a log CMOS image sensor: (a) basic pixel circuit; (b) circuit including accumulation mode. Illustrated after [179].

This current I_c is charged to the capacitor C and thus the time variation of V_{out} is given by

$$C\frac{dV_{out}}{dt} = I_c. \tag{3.10}$$

By substituting Eq. 3.8 into Eq. 3.9, we obtain the following equation:

$$I_c = I_{ph}\exp\left[\frac{e}{mk_BT}\left(V_{out} - V_{ps}\right)\right]. \tag{3.11}$$

By substituting this into Eq. 3.10 and integrating, the output voltage V_{out} is obtained as

$$V_{out} = \frac{mkT}{e}\ln\left(\frac{e}{mkTC}\int I_{ph}dt\right) + V_{ps}. \tag{3.12}$$

Although a log sensor has a wide dynamic range over 100 dB, it has some disadvantages, such as low photo-sensitivity especially in the low illumination region compared with a 4T-APS; slow response and a relatively large variation of the device characteristics owing to sub-threshold operation.

3.2.1.3 Capacitive Transimpedance Amplifier pixel

Conventional APS has a source follower (SF) amplifier in a pixel, although its current load is placed in a column. The gain of a SF amplifier is about 1. To increase the sensitivity, capacitive transimpedance amplifier (CTIA) can be integrated in a pixel [180–183]. In the past, some MOS imagers integrated the CTIA as a column amplifier [67]. Advanced CMOS technology makes it possible to integrate the CTIA in a pixel [181–183].

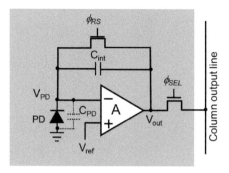

FIGURE 3.4: Pixel circuit employed with a capacitive transimpedance amplifier in a pixel. Illustrated after [182].

In Fig. 3.4, at the PD node, considering the total current is zero, the following equation holds,

$$I_{ph} + C_{PD}\frac{dV_{PD}}{dt} + C_{int}\frac{d(V_{PD} - V_{out})}{dt} = 0, \qquad (3.13)$$

where I_{ph} is the photocurrent of the PD. By using Eq. 3.13, the output voltage V_{out} is expressed as [181]:

$$\frac{dV_{out}}{dt} = \frac{I_{ph}}{C_{int}\left(1 - \left(1 + \frac{1}{A}\frac{C_{PD}}{C_{int}}\right)\right)},$$

$$\therefore V_{out} = \frac{1}{C_{int}}\int I_{ph}dt, \qquad (3.14)$$

$$\text{If } A \gg \frac{C_{PD}}{C_{int}} > 1,$$

where A is an open loop gain of the amplifier and $V_{out} = AV_{PD}$. By Eq. 3.14, the output voltage is not dependent on the photodiode capacitance C_{PD} in 3T-APS, and instead, can be controlled by the capacitance C_{int}.

3.2.1.4 Lock-in pixel

The photo-generated carriers in a photodiode of an image sensor are typically moved by diffusion, because there is no electric field in the in-plane direction of a photodiode. In a lock-in pixel, an electric field is introduced in the in-plane direction of the photodiode so as to drift the photo-generated carriers and reach the electrode quickly. If the electrodes are placed facing each other in a photodiode, the photo-generated carriers can be modulated by applying voltages to the two electrodes. Such function can be used for lock-in operation, where the signal by photo-generated carriers can be modulated by the external modulation signal. The structure of a typical lock-in pixel is shown in Fig. 3.5 [184].

The lock-in pixel is typically used for demodulation device and TOF (Time-of-Flight) imager, which are described in Sec. 4.5 and Sec. 4.6 of Chapter

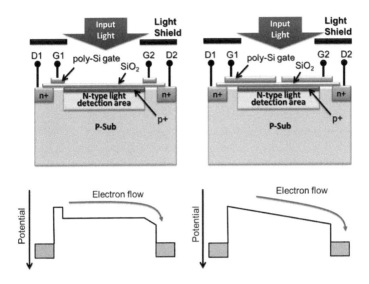

FIGURE 3.5: Schematic structure of lock-in pixel. Illustrated after [184].

4, respectively. Several types of lock-in pixels have been developed, such as PMD (photo mixer device) [185], CAPD (current-assisted photonic demodulator) [186, 187], LEFM (lateral electric field charge modulator) [188], etc. It is important for them to achieve a high SNR with an affordable spatial resolution (or pixel numbers) as well as lock-in pixel function.

3.2.2 Pulse modulation mode

While in an APS the output signal value is read after a certain time has passed, in pulse modulation (PM) the output signal is produced when the signal reaches a certain value. The sensors, which use PM are called PM sensors, time-to-saturation sensors [189], and address event representation sensors [56]. The basic structure of pulse width modulation (PWM) and pulse frequency modulation (PFM) are shown in Fig. 3.6. Other pulse schemes such as pulse amplitude modulation and pulse phase modulation are rarely used in smart CMOS image sensors.

The concept of a PFM-based photo-sensor was first proposed by K.P. Frohmader [190] and its application to image sensing was first reported by K. Tanaka *et al.* [106], where a GaAs MSM photodetector was used to demonstrate the fundamental operation of the sensor. MSM photodetectors are discussed in Sec. 2.3.5. A PWM-based photo-sensor was first proposed by R. Müller [191] and its application to an image sensor was first demonstrated by V. Brajovic and T. Kanade [192].

PM has the following features:

- Asynchronous operation
- Digital output

- Low voltage operation

Because each pixel in a PM sensor can individually make a decision to output, a PM sensor can operate without a clock, i.e., asynchronously. This feature provides adaptive characteristics for ambient illuminance with a PM-based image sensor and thereby allows application to wide dynamic range image sensors.

Another important feature is that a PM sensor acts as an ADC. In PWM, the count value of the pulse width is a digital value. An example of a PWM is shown in Fig. 3.6, which is essentially equivalent to a single slope type ADC. The PFM is equivalent to a one-bit ADC.

In a PM-based sensor, the output is digital, and hence, is suitable for low-voltage operation. In the following sections, some example PM image sensors are described.

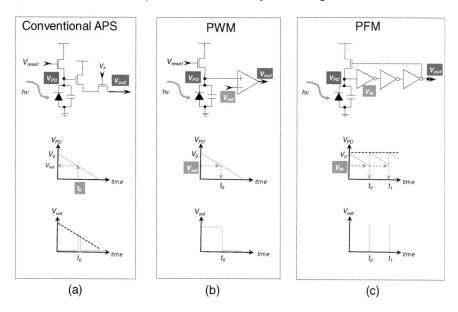

FIGURE 3.6: Basic circuits of pulse modulation. (a) conventional 3T-APS. (b) pulse width modulation (PWM). (c) pulse frequency modulation (PFM).

3.2.2.1   Pulse width modulation

R. Müller first proposed and demonstrated an image sensor based on PWM [191]. Subsequently, V. Brajovic and T. Kanade proposed and demonstrated an image sensor using a PWM-based photo-sensor [192]. In this sensor, circuits are added to calculate a global operation summing the number of on-state pixels, allowing cumulative evolution to be obtained in an intensity histogram.

The digital output scheme is suitable for on-chip signal processing. M. Nagata *et al.* have proposed and demonstrated a time-domain processing scheme using PWM and that PWM is applicable to low-voltage and low-power design in deep sub-micron technology [193]. They have also demonstrated a PWM-based image sensor that realizes on-chip signal processing of block averaging and 2D-projection [194].

The low-voltage operation feature of PWM has been demonstrated in Refs. [195, 196], where a PWM-based image sensor was operated under a 1-V power supply voltage. In particular, S. Shishido *et al.* [196] have demonstrated a PWM-based image sensor with pixels consisting of three transistors plus a PD. This design overcomes the disadvantage of conventional PWM-based image sensors of requiring a number of transistors for a comparator.

PWM can be applied to enhance a sensor's dynamic range, as described in Sec. 4.4.3.3, and much research on this topic has been published. Several advantages of this use of PWM, including the improved DR and SNR of PWM, are discussed in Ref. [189].

PWM is also used as a pixel-level ADC in digital pixel sensors (DPSs) [64, 197, 198]. Some sensors use a simple inverter as a comparator to minimize the area of a pixel so that a processing element can be employed in the pixels [64]. W. Bidermann *et al.* have implemented a conventional comparator and memory in a chip [198]. In Fig. 3.7(b), a ramp waveform is input to the comparator reference terminal. The circuit is almost the same as a single slope ADC. This type of PWM operates synchronously with the ramp waveform.

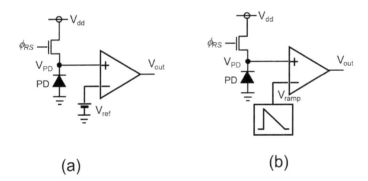

(a)   (b)

FIGURE 3.7: Basic circuits of pulse width modulation (PWM)-based photo-sensors; two types of PWM photo-sensors are illustrated: one uses a fixed threshold for a comparator (a) and the other uses a ramp waveform for a comparator (b).

3.2.2.2 Pulse frequency modulation

PWM produces an output signal when the accumulation signal reaches a threshold value. In pulse frequency modulation (PFM), when the accumulation signal reaches the threshold value, the output signal is produced, the accumulated charges are reset, and the accumulation starts again. As this process is repeated, the output signals continue to be produced. The frequency of the output signal production is proportional to the input light intensity. PFM-like coding systems are found in biological systems [199], which have inspired the pulsed signal processing [200, 201]. K. Kagawa *et al.* have also developed the pulsed image processing [202], described in Chapter 5. T. Hammadou has discussed stochastic arithmetic in PFM [203]. A PFM-based image sensor was first proposed by K. Tanaka *et al.* [106] and demonstrated by W. Yang [204] for a wide dynamic range, with additional details given in Refs. [156, 205, 206].

One application of PFM is the address event representation (AER) [207], which is applied, for example, in sensor network camera systems [208, 209].

PFM-based photo-sensors are used in biomedical applications, such as in ultra-low light detection in biotechnology [210, 211], which is describe in Sec. 4.2.1.1 of Chapter 4. Another application of PFM in the biomedical field is retinal prosthesis. The application of PFM photo-sensors to the retinal prosthesis of sub-retinal implantation was first proposed in Ref. [212] and has been continuously developed by the same group [202, 205, 213–223], as well as by other groups [224–227]. The retinal prosthesis is described in Sec. 5.5.2.

3.2.2.2.1   Operational principle of PFM   The operational principle of PFM is as follows. Figure 3.8 shows a basic circuit of a PFM photo-sensor cell. From the circuit, the sum of the photocurrent I_{ph} including the dark current I_d discharges the PD capacitance C_{PD}, which is charged to V_{dd}, thus causing V_{PD} to decrease. When V_{PD} reaches the threshold voltage V_{th} of the inverter, the inverter chain is turned on and an output pulse is produced. The output frequency f is approximately expressed as

$$f \approx \frac{I_{ph}}{C_{PD}(V_{dd} - V_{th})}. \tag{3.15}$$

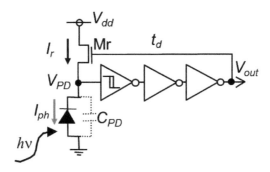

FIGURE 3.8: Basic circuits of a PFM photo-sensor.

Figure 3.9 shows the experimental results for the PFM photo-sensor presented in Fig. 3.8. The output frequency increases in proportion to the input light intensity. The dynamic range is measured to be near 100 dB. In the low light intensity region, the frequency saturates owing to the dark current.

The inverter chain including the Schmitt trigger has a delay of t_d and the reset current I_r provided by the reset transistor M_r has a finite value. The following is

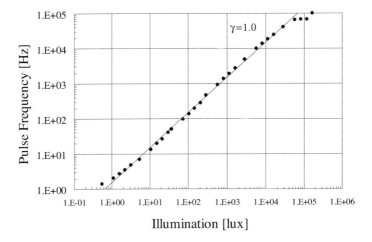

FIGURE 3.9: Experimental output pulse frequency dependence on the input light intensity for the PFM photo-sensor presented in Fig. 3.8.

an analysis considering these parameters, while a more detailed analysis appears in Ref. [217]. The PD is discharged by the photocurrent I_{ph}; and is charged by the reset current I_r minus I_{ph}, because a photocurrent is still generated during the charging.

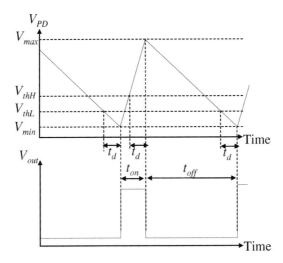

FIGURE 3.10: Time course of V_{PD}, taking the delay t_d into consideration.

By taking t_d and I_r into consideration, V_{PD} varies as in Fig. 3.10. From the figure, the maximum voltage V_{max} and minimum voltage V_{min} at V_{PD} are expressed as

$$V_{max} = V_{thH} + \frac{t_d(I_r - I_{ph})}{C_{PD}},$$

(3.16)

$$V_{min} = V_{thL} - \frac{t_d I_{ph}}{C_{PD}}.$$

(3.17)

Here, V_{thH} and V_{thL} are the upper and lower thresholds of the Schmidt trigger. It is noted that the discharging current is I_{ph}, while the charging current or reset current is $I_r - I_{ph}$. t_{on} and t_{off} in Fig. 3.10 are given by

$$\begin{aligned} t_{on} &= \frac{C_{PD}(V_{thH} - V_{min})}{I_r - I_{ph}} + t_d \\ &= \frac{C_{PD}V_{th} + t_d I_r}{I_r - I_{ph}}, \end{aligned}$$

(3.18)

$$\begin{aligned} t_{off} &= \frac{C_{PD}(V_{max} - V_{thL})}{I_{ph}} + t_d \\ &= \frac{C_{PD}V_{th} + t_d I_r}{I_{ph}}, \end{aligned}$$

(3.19)

where $V_{th} = V_{thH} - V_{thL}$. t_{on} is the time when the reset transistor M_r charges the PD, i.e., when M_r turns on. During this time, the pulse is on-state and hence it is equal to the pulse width. t_{off} is the time when M_r turns off. During this time the pulse is in the off-state. The pulse frequency f of the PFM photo-sensor is expressed as

$$\begin{aligned} f &= \frac{1}{t_{on} + t_{off}} \\ &= \frac{I_{ph}(I_r - I_{ph})}{I_r(C_{PD}V_{th} + t_d I_r)} \\ &= \frac{I_r^2/4 - (I_{ph} - I_r/2)^2}{I_r(C_{PD}V_{th} + t_d I_r)}. \end{aligned}$$

(3.20)

If the reset current of Mr I_r is much larger than the photocurrent I_{ph}, then Eq. 3.20 becomes

$$f \approx \frac{I_{ph}}{C_{PD}V_{th} + t_d I_r}.$$

(3.21)

Thus, the pulse frequency f is proportional to the photocurrent I_{ph}, i.e., the input light intensity.

In addition, Eq. 3.20 shows that the frequency f becomes maximum at a photocurrent of $I_r/2$, and then decreases. Its maximum frequency f_{max} is

$$f_{max} = \frac{I_r}{4(C_{PD} + t_d I_r)}.$$

(3.22)

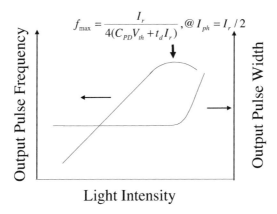

$$f_{max} = \frac{I_r}{4(C_{PD}V_{th} + t_d I_r)}, @ I_{ph} = I_r/2$$

Output Pulse Frequency

Output Pulse Width

Light Intensity

FIGURE 3.11: Pulse frequency and pulse width dependence on the input light intensity.

The pulse width τ is

$$\tau = t_{on} = \frac{C_{PD}V_t h + t_d I_r}{I_r - I_{ph}}. \tag{3.23}$$

Figure 3.11 shows the pulse frequency and pulse width dependence on the input light intensity based on the above equations, 3.22 and 3.23.

Self-reset operation   Under strong light conditions, it is difficult to detect weak light owing to the saturation of the photodiode. To solve this problem, a self-reset type CMOS image sensor has been developed [228, 229].

The principle of the self-reset type CMOS image sensor is almost the same as that of PFM. When the light is very strong, the signal increases and finally saturates due to the limited well capacity/stored charge. Consequently, the signal over this intensity level cannot be measured, and the shot noise also increases. For a self-reset type image sensor, if the light causes well-saturation, the photodiode resetting automatically occurs by using a feedback path similar to PFM.

The self-reset type image sensor has been implemented as shown in Fig. 5.51 [229]. The number of transistors per pixel is only 11, because reset counting circuits are not implemented in a pixel. Table 3.2 shows the comparison in the specification of previously reported PFM sensors.

In PFM image sensors, the reset number is essential for constructing the image, while in self-reset type image sensors, the output data after the final reset action is used for the image. Figure 3.14 clearly demonstrates an increase in SNR of over 10 dB by using the self-reset mode. The application of a self-reset type image sensor to measure the neural signals in a rat brain is described in Sec. 5.4.3.1.4 of Chapter 5.

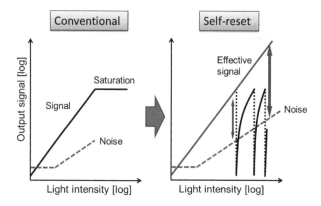

FIGURE 3.12: The principle of the self-reset type image sensor compared with a conventional image sensor.

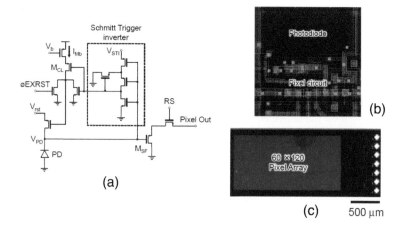

FIGURE 3.13: The self-reset type CMOS image sensor: (a) pixel circuits, (b) pixel layout, and (c) chip photo. Adapted from [229] with permission.

3.2.2.3 SPAD

As introduced in Chapter 2, a single photon avalanche diode (SPAD) is an avalanche photodiode working in Geiger mode, and outputs a pulse train. SPAD is now described in this section. Figure 3.15 (a) shows the pixel structure of an APD in standard CMOS technology. The APD is fabricated in a deep n-well and multiple regions are surrounded by a p^{+}-guard ring region. In the Geiger mode, the APD produces a spike-like signal, and not an analog output, that is a SPAD. A pulse shaper with an inverter is used to convert the signal to a digital pulse, as shown in Fig. 3.15. There are two modes for SPAD: the photon counting mode and the photon timing

TABLE 3.2
Comparison of PFM based image sensors. COMP: comparator, ST: Schmitt trigger.

Reference	[230]	[231]	[232]	[233]	[229]
Technology	0.25 μm	0.35 μm	0.18 μm	0.50 μm	0.35 μm
Number of Trs. / pixel	N/A	N/A	43	28	11
Pixel size [μ m^2]	45 × 45	25 × 25	19 × 19	49 × 49	15 × 15
Self-reset type	COMP	COMP	COMP	ST	ST
Counter [bit]	8	1	6	6	-
Fill factor [%]	23	27	50	25	26
Circuit area [μ m^2 × 10^2]	16	4.6	18	18	1.7
Max. frame rate [kHz]	1	0.015	1	> 1	0.3
Peak SNR [dB]	N/A	74.5	55.6	65	64

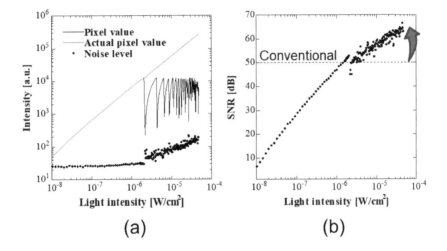

(a)   (b)

FIGURE 3.14: Experimental results of the self-reset type CMOS image sensor. (a) Value of output signal and noise signal as a function of light intensity. (b) SNR as a function of light intensity. Adapted from [229] with permission.

mode [234]. In the photon counting mode, it is mainly used to measure the intensity of slowly varying but very weak optical signals in the micro-second range, and in the photon timing mode it is used for reconstructing a very fast optical waveform in the pico-second range [234]. In the photon counting mode, the number of output pulses is counted in a certain time slot so as to obtain the intensity change with time. In the photon timing mode, time-to-digital counter (TDC) can be integrated in a SPAD pixel or on a chip. In both the modes, averaging operations are required.

Recently, a 512 × 512-pixel SPAD image sensor has been published [235]. There are 11 transistors in a pixel, with an avalanche photodiode. The sensor is composed of a SPAD, a 1-bit memory, and a gating mechanism capable of turning the SPAD on and off.

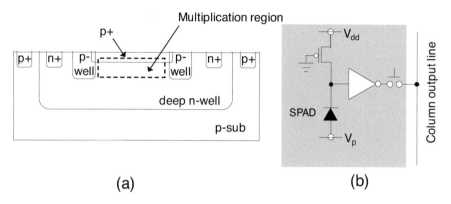

(a)   (b)

FIGURE 3.15: Basic structure of (a) an APD in standard CMOS technology and (b) CMOS circuits for the Geiger mode APD in a pixel. The PMOS connected to V_{dd}, acts as a resistor for the quenching process. V_p is a negative voltage value that forces the PD into the avalanche breakdown region. SPAD: single photon avalanche diode. Illustrated after [93].

3.2.3 Digital mode

Digital processing architecture in a smart CMOS image sensor is based on the concept of employing an ADC in each pixel as shown in Fig. 3.16 [63, 64]. In some cases, the ADCs are placed in a column parallel manner [236]. They are called vision chips, wherein the processing is done on a chip. CMOS image sensors with digital outputs are often called digital pixel sensors (DPSs) [197, 198]. Thanks to digital signal output, a DPS has fast response speed. In addition, it is possible to realize the programmability on a chip.

   Figure 3.16 shows the fundamental circuits of a digital pixel sensor. It consists of a photodetector, a buffer, and an ADC. Because generally, the ADCs are area consuming circuits, implementing an ADC with a small area size in a pixel is a key issue.

   As mentioned above, a key feature of the digital processing architecture is the implementation of an ADC in a pixel or in-pixel ADC. The area efficiency is one of the most important factors for in-pixel ADCs. In the comparison table in Table 2.3 in Sec. 2.6.3 of Chapter 2, a single slope (SS) ADC (See Sec. 2.6.3 of Chapter 2) is a candidate for in-pixel ADC, and in Ref. [197], an SS ADC with an 8-bit memory is employed in a pixel as shown in Fig. 3.17.

   DPSs are very attractive for smart CMOS image sensors because they are programmable with a definite precision.   In addition, DPSs have achieved

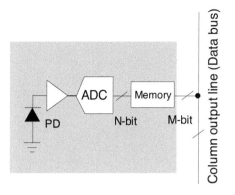

FIGURE 3.16: Fundamental pixel circuits of a digital pixel sensor.

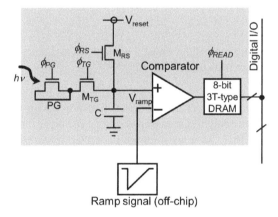

FIGURE 3.17: A DPS with a single slope ADC in a pixel. Illustrated after [197].

high-speeds of 10 Kfps [197] and a wide dynamic range of over 100 dB [198]. Thus it is suitable for robot vision, which requires versatile and autonomous operation with a fast response.

There are simpler ADCs compared with those in Table 2.3. Employing an ADC in a pixel enables programmable operation with fast processing speed. In Ref. [63], a simple PWM scheme is introduced with an inverter for a comparator. Further, the nearest neighboring operation can be achieved using digital architecture. Figure 3.18 shows a pixel block diagram reproduced from Ref. [65]. It is noted that this sensor requires no scanning circuits, because each pixel transfers digital data to the next pixel. This is another feature of the fully programmable digital processing architecture. In Ref. [237], a pixel-level ADC was effectively used to control the conversion curve or gamma value. The programmability of the sensor is utilized to enhance the captured image, as in logarithmic conversion, histogram equalization,

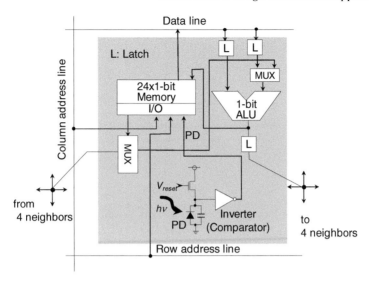

FIGURE 3.18: Pixel circuit diagram of a smart CMOS image sensor with digital processing architecture, Illustrated after [65].

and similar techniques. The challenge for digital processing technology is pixel resolution, which is currently restricted by the large number of transistors in a pixel; for example, in Ref. [63], 84 transistors are integrated in each pixel with a size of 80 μm $\times$ 80 μm using a 0.5-μm standard CMOS technology. By using a finer CMOS technology, it would be possible to make a smaller pixel with a lower power consumption and higher processing speed. However, such a fine technology has problems such as low voltage swing, low photo-sensitivity, etc. In some digital pixel sensors, an ADC is shared by four pixels [197, 198]. Recently, the stacked CMOS image sensor technology described in Sec. 3.3.4 has emerged, which can alleviate the issue of how to integrate ADC in a pixel. In the near future, many smart CMOS image sensors will be DPSs with stacked technology.

3.3   Smart materials and structures

In this section, several materials other than single crystal silicon are introduced for smart CMOS image sensors. As described in Chapter 2, the visible light is absorbed in silicon, which means that silicon is opaque to visible wavelengths. Some materials are transparent in the visible wavelength region, such as SiO_2 and sapphire (Al_2O_3), which are introduced in modern CMOS technology as SOI (silicon-on-insulator) and SOS (silicon-on-sapphire). The detectable wavelength of silicon is determined by its bandgap, which corresponds to a wavelength of about 1.1 μm. Other materials, such as SiGe and Ge can respond to a longer wavelength light than silicon. The fundamental characteristics of Ge are summarized in Appendix A.

3.3.1 Silicon-on-insulator

Recently, SOI CMOS technology has been developed for low voltage circuits [238]. The structure of SOI is shown in Fig. 3.19, in which a thin Si layer is placed on a buried oxide (BOX) layer. The top Si layer is placed on a SiO_2 layer or an insulator. A conventional CMOS transistor is called a bulk MOS transistor to clearly distinguish it from an SOI MOS transistor. MOS transistors are fabricated on an SOI layer and are completely isolated via shallow trench isolation (STI), which penetrates the BOX layer, as shown in Fig. 3.19(b). These transistors exhibit lower power consumption, less latch-up, and less parasitic capacitance compared to the bulk CMOS technology, in addition to other advantages [238].

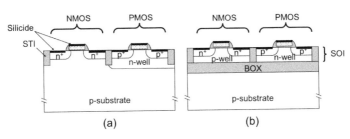

FIGURE 3.19: Cross-section of (a) bulk and (b) SOI CMOS devices. STI: shallow trench isolation, SOI: silicon-on-insulator, BOX: buried oxide. Silicide is a compound material with silicon and metal such as $TiSi_2$.

SOI technology is attractive for CMOS image sensors for the following reasons:

- SOI technology can produce circuits with low voltage and low power consumption [239]. For mobile applications, sensor networks, and implantable medical devices, this feature is important.
- When using SOI processes, NMOS and PMOS transistors can be employed without sacrificing the area, compared to the bulk CMOS processes where an N-well layer is necessary to construct PMOS transistors on a p-type substrate. Figure 3.19, which compares the bulk and SOI CMOS process, clearly illustrates this. A PMOSFET reset transistor is preferable for use in an APS due to the fact that it exhibits no voltage drop, in contrast to an NMOSFET reset transistor.
- SOI technology makes it easy to fabricate a BSI image sensor, discussed in the following Sec. 3.3.3.
- The SOI structure is useful for preventing crosstalk between pixels caused by diffused carriers; photo-carriers generated in a substrate can reach the pixels in the SOI image sensor. In SOI, each pixel is isolated electrically.

- The SOI technology can also be used in 3D integration [240, 241]. A pioneering work on the use of SOI in an image sensor for 3D integration appears in Ref. [240]. By introducing damascened Au electrodes, direct bonding method can achieve pixel-wise electrical interconnection [242]. The 3D integration for image sensors is described in Sec. 3.3.4.

Since an SOI layer is generally so thin (usually below 200 nm) that the photo-sensitivity is degraded. To obtain a good photo-sensitivity, several methods have been developed. The most compatible method with conventional APSs is to create a PD region on a substrate [243]. This ensures that the photo-sensitivity is the same as that of a conventional PD. However, it requires modifying the standard SOI fabrication process. Post-processing for the surface treatment is also important to obtain low dark current. The second method is to employ a lateral PTr, as shown in Fig. 2.6 (d) in Sec. 2.3 [244]. As the lateral PTr has a gain, the photo-sensitivity increases even in a thin photo-detection layer. Another application of SOI is in lateral pin PDs [245], although in this case the photo-detection area and the pixel density is a trade-off. A PFM photo-sensor, described in Sec. 3.2.2.2, is especially effective as an SOS imager, which is discussed later in this section.

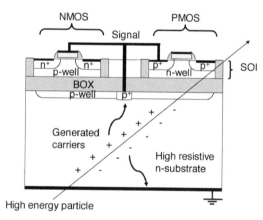

FIGURE 3.20: Cross-section of SOI detector for high energy particle. It consists of thick high resistive n-type substrate, BOX, and SOI. The pn junction is formed inside the substrate and the p-side is connected to the SOI circuits with a VIA through the BOX layer from the substrate to SOI. Illustrated after [246].

SOI is widely used in micro-electro-mechanical systems (MEMS), as a beam structure of silicon can easily be fabricated by etching a BOX layer with a selective SiO_2 etchant. An application of such a structure in an image sensor is an uncooled focal plane array (FPA) infrared image sensor [247]. The IR detection is achieved with a thermally isolated pn-junction diode. The thermal radiation moves the pn-junction built-in potential and by sensing this shift the temperature can be measured and thus IR radiation is detected. By combining MEMS structures, the range of potential applications of SOI for image sensors will be numerous.

Recently, SOI based CMOS image sensors are developed for the detection of high energy particles [246, 248]. Figure 3.20 illustrates a schematic cross-sectional view of a high energy particle detector based on SOI structure [246]. The incident high energy particle produces electron-hole pairs inside the thick high resistive substrate. In the figure, the generated holes are detected in the pn junction under the BOX layer. The pn diode is connected to the SOI circuits for detecting the signal.

3.3.1.1 Silicon-on-sapphire

Silicon-on-sapphire (SOS) is a technology using sapphire as a substrate instead of silicon [249, 250]. A thin silicon layer is directly formed on a sapphire substrate. It is noted that the top silicon layer is neither poly- nor amorphous silicon but a single crystal of silicon, and thus the physical properties, such as mobility, are almost the same as in an ordinary Si-MOSFET. Sapphire is Al_2O_3. It is transparent in the visible wavelength region and hence image sensors using SOS technology can be used as backside illumination sensors without any thinning process [214, 245, 251, 252]; however, some polishing is required to make the back surface flat. Lateral PTrs were used in Ref. [245], while a PFM photo-sensor is used in the work of Ref. [214, 251] due to the low photo-sensitivity in a thin detection layer. Figure 3.21 shows an image sensor fabricated by SOS CMOS technology. The chip is placed on a sheet of printed paper and the printed pattern on the paper can be seen through the transparent substrate.

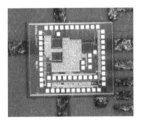

FIGURE 3.21: PFM photo-sensor fabricated using SOS technology.

3.3.2 Extending to NIR region

Usually silicon has a sensitivity of up to 1.1 μm, determined by the bandgap of silicon $E_g(\text{Si}) = 1.12$ eV. Even in the wavelength shorter than 1.1 μm, it is generally difficult to detect the NIR light with the Si-based photodiode. In this section, to enhance the NIR region, two methods are mentioned. Here, NIR is defined as the light at the wavelength between 0.75–1.4 μm (See Appendix D.1). The first one is to introduce a multi-path for the NIR light inside Si. Even with a small absorption coefficient in the NIR region, such a multi-path extends the total traveling distance for photons. The second one is to introduce materials other than silicon.

3.3.2.1 Multi-path structure

By introducing a number of fine inverted pyramidal structures in the surface of silicon, the incident light path can be extended as shown in Fig. 3.22 [253]. Since the incident light travels a long distance in such a multi-path structure, the total amount of absorption of the NIR light increases so as to become detectable even if the absorption in the NIR region is small . In Ref. [253], the sensitivity at 850 nm increases 80% compared with the sensor with a flat surface.

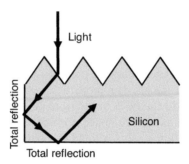

FIGURE 3.22: Fine inverted pyramidal structure enables multi-paths of incident light and extend the detectable NIR region close to the bandgap wavelength of Si. Illustrated after [253].

The other method to extend the light traveling path is to introduce black Si [254]. Black Si is a nano-structure. Almost no reflection occurs in the visible to NIR range, and it appears black. The fabrication process of black silicon is compatible to that of conventional CMOS image sensors. The advantage of black silicon is its easy fabrication process compared with the inverted pyramidal structure. On the contrary, its disadvantage is that the photo-sensitivity may not be uniform for an image sensor with small-size pixels.

3.3.2.2 Materials other than silicon

To extend the sensitivity beyond 1.1 μm, materials other than conventional silicon crystals must be used. There are many materials with a sensitivity at longer wavelengths than that of silicon. To realize a smart CMOS image sensor with a sensitivity at longer wavelengths than silicon, hybrid integration of materials with a longer wavelength photo-sensitivity, such as SiGe, Ge, InGaAs, InSb, HgCdTe, PbS, and quantum-well infrared photodetector (QWIP) [255], as well as some others [256] is recommended. The wavelength where these materials has photo-sensitivity is called short-wavelength infrared (SWIR) (1.4–3 μm, See Appendix D.1). In addition to black Si and SiGe, these materials can be placed on a silicon readout integrated circuit (ROIC) bonded by flip-chip bonding through metal bumps. Several methods to realize SWIR detection using ROIC are given in Ref. [257]. For example, a SWIR image sensor with PbS quantum dot film placed on ROIC has been developed [258].

Over 3 μm region or mid-wavelength infrared (MWIR) (3–8 μm, See Appendix D.1), Schottky barrier photodetectors such as PtSi (platinum silicide) are widely used in IR imagers [259], which can be monolithically integrated on a silicon substrate. These MWIR image sensors usually work under low temperature conditions, but recently uncooled type MWIR/LWIR image sensors have been developed. There have been many reports of MWIR/LWIR image sensors with this configuration, and it is beyond the scope of this book to introduce them.

Before describing the sensors, we briefly overview the materials SiGe and germanium. Si_xGe_{1-x} is a mixed crystal of Si and Ge with an arbitrary composition x [260]. The bandgap can be varied from that of Si ($x = 1$), $E_g(\text{Si}) = 1.12$ eV or $\lambda_g(\text{Si}) = 1.1$ μm to that of Ge ($x = 0$), $E_g(\text{Ge}) = 0.66$ eV or $\lambda_g(\text{Ge}) = 1.88$ μm. SiGe on silicon is used for hetero-structure bipolar transistors (HBT) or strained MOS FETs in high-speed circuits. The lattice mismatch between the lattice constants of Si and Ge is so large that it is difficult to grow a thick SiGe epitaxial layers on the Si substrate. Recently, Ge on Si technology has been advanced for high speed receivers in optical fiber communication where light at 1.3-1.5 μm wavelength region is used [261]. The high quality of epitaxial Ge layer on Si has been obtained by various methods to alleviate the large lattice mismatch between Si and Ge [261]. By using these technologies developed for high speed optical communication, image sensors with the Ge detection layers have been developed [262, 263]. The other material in NIR – SWIR is InGaAs.

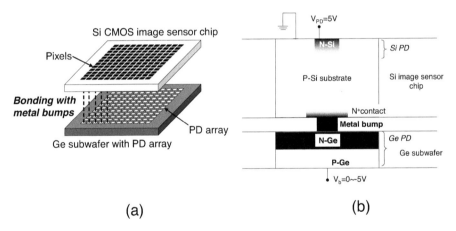

(a)   (b)

FIGURE 3.23: Smart CMOS image sensor that can detect in both the visible and eye-safe regions: (a) chip structure, (b) cross-section of the sensor. Adapted from [264] with permission.

Here, we introduce one example of a smart CMOS image sensor with sensitivities in both the visible and NIR regions, called the eye-safe wavelength region [264,265]. The human eye is more tolerant to the eye-safe wavelength region (1.4–2.0 μm) than the visible region because more light in the eye-safe region is absorbed at the cornea than light in the visible region and thus less damage is caused to the retina.

The sensor consists of a conventional Si-CMOS image sensor and a Ge PD array formed underneath the CMOS image sensor. The capability of the sensor to capture visible images is not affected by extending its range into the IR region. The operating principle of the NIR detection is based on photo-generated carrier injection into a Si substrate from a Ge PD. The structure of the device is shown in Fig. 3.23.

Photo-generated carriers in the Ge PD region are injected into the Si substrate and reach the photo-conversion region at a pixel in the CMOS image sensor by diffusion. When the bias voltage is applied, the responsivity in the NIR region increases, as shown in Fig. 3.24. The inset of the figure shows a test device for the experiment of the photo-response. It is noted that the NIR light can be detected at the Ge PD placed at the back of the sensor because the silicon substrate is transparent in the NIR wavelength region.

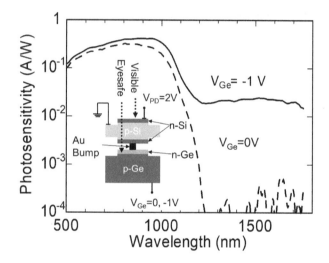

FIGURE 3.24: Photosensitivity curve as a function of the input light wavelength. The bias voltage V_b of the Ge PD is a parameter. The inset shows the test structure for this measurement. Adapted from [264] with permission.

3.3.3 Backside illumination

As shown in Fig. 3.25, a backside illumination (BSI) CMOS image sensor has advantages, such as a large FF and a large optical response angle compared to a conventional CMOS image sensor or front-side illumination (FSI) CMOS image sensor [266–270]. Figure 3.25(a) shows a cross-section of a conventional CMOS image sensor or FSI CMOS image sensor, where the input light travels a long distance from the micro-lens to the PD, which causes crosstalk between pixels. In addition, the metal wires form obstacles for the light. In a BSI CMOS image sensor, the distance between the micro-lens and the PD can be reduced so that the optical characteristics are significantly improved. As the p-Si layer on the PD must be thin to reduce the absorption in the layer as much as possible, the substrate is typically ground to be thin. It is noted that this structure is similar to that of a CCD where a very thin SiO_2 layer exists between the PD and metal layer to minimize the optical crosstalk.

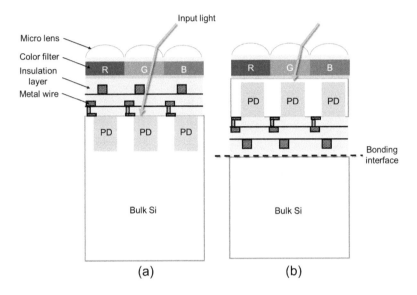

FIGURE 3.25: Cross-section of (a) a conventional CMOS image sensor and (b) a backside illumination CMOS image sensor.

3.3.4   3D integration

Three-dimensional (3D) integration has been developed to integrate more circuits in a limited area. An image sensor with a 3D integration structure has an imaging area on its top surface and signal processing circuits in the subsequent layers. The 3D integration technology thus makes it easy to realize pixel-level processing or pixel-parallel processing.

Recently, based on the structure of backside illumination (BSI) CMOS image sensor, 3D integrated or stacked CMOS image sensors have been developed. Figure 3.26 shows a conceptual illustration of a stacked image sensor chip and its cross-sectional structure. The 3D image sensor in the figure is fabricated after the fabrication of the backside-illuminated CMOS image sensor (CIS) wafer, which is bonded with another SOI based wafer where circuits are employed.

There are two types to realize 3D integration; one is to use TSV (through silicon via) and the other is to use micro-bumps to bond two wafers directly through the bumps. In the TSV technology, after two wafers are bonded, TSVs are formed for electrical interconnection between the two wafers. Usually, the TSVs are formed in the peripheral of the chip [271, 272] as shown in Fig. 3.27(a). On the contrary, the micro-bump method can realize pixel-wise interconnections or pixel block interconnections [273–275] as shown in Fig. 3.27(b).

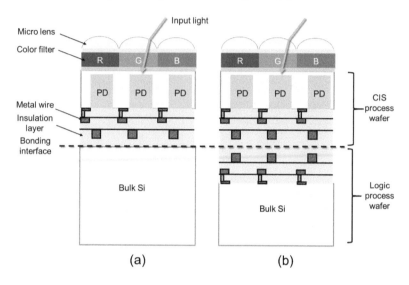

FIGURE 3.26: 3D image sensor chip; (a) Backside-illuminated CMOS sensor chip that is fabricated on a CIS (CMOS image sensor) process wafer as shown in Fig. 3.25(b); (b) 3D CMOS image sensor where CIS process wafer and logic process wafer are bonded.

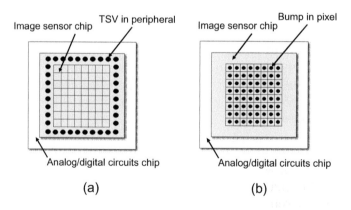

FIGURE 3.27: Two methods to realize 3D image sensor chip: (a) TSV method, wherein the TSVs are formed in the peripheral of the chip; (b) Micro-bump method, in which micro-bumps are formed in each pixel so as to realize pixel-wise interconnection.

Using the TSV method, three different process wafers, BSI-CIS, DRAM and logic process wafers, are stacked [271, 272]. The scanning electron microscope (SEM) cross-sectional view is shown in Fig. 3.28.

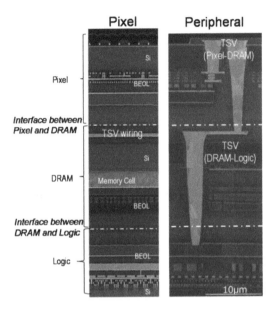

FIGURE 3.28: A SEM cross-sectional view of a CMOS image sensor with 3-stacked wafers. In the chip, the top part, the middle part, and the bottom part consist of BIS-CIS process wafer, DRAM process wafer and logic process wafer, respectively. Adapted from [272] with permission.

3.3.5 Smart structure for color detection

Usually, image sensors can detect colorsignals and can separate light into elementary color signals, such as RGB by using on-chip color filters . Conventional methods for color realization are described in Sec. 2.8. Other methods to realize color using smart functions are summarized in the following subsections.

3.3.5.1 Stacked organic PC films

The first introduced method to acquire RGB-color used three stacked organic photo-conductive (OPC) films that can be fabricated on a pixel [102–104, 276] as shown in Fig. 3.29(a). Each of the organic films acts as a PC detector (see Sec. 2.3.5) and produces photocurrents according to its light sensitivity. This method can almost realize a 100% FF. Connecting the stacked layers is a key issue here.

Another organic image sensor uses an organic film for light detection instead of Si and forms conventional on-chip RGB color filters [277, 278]. It is noted that the

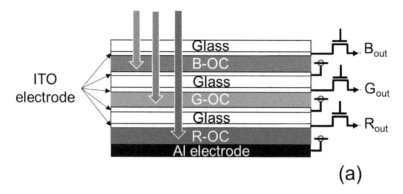

(a)

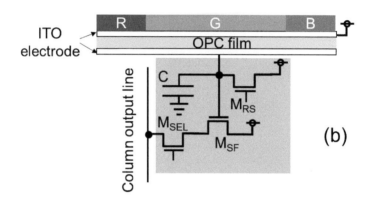

(b)

FIGURE 3.29: Device structure of the organic image sensors. (a) Three organic films are used corresponding to blue, green and red color detection. Illustrated after [276]. (b) One organic photo-conductive (OPC) film is used for light detection layer. The color is realized by conventional on-chip color filters. Illustrated after [277].

absorption coefficient of organic photo-conductive film is one order of magnitude larger than that of Si. Thus the thickness of the OPC film can be thinner than that of a photodiode in the conventional CMOS image sensor. This method can also almost realize a 100% FF. In Fig. 3.29(b), the photo-conductive voltage can be sensed by the source follower transistor M_{SF}. The reset transition and select transistor are M_{RS} and M_{SEL}, respectively. The capacitor is incorporated to extend the saturation level of the pixel [277].

3.3.5.2 Multiple junctions

The photo-sensitivity in silicon depends on the depth of the pn-junction. Thus, having two or three junctions located along a vertical line alters the photo-sensitivity spectrum [279–281]. To adjust the three junction depths, the maximum photo-sensitivities corresponding to the RGB colors are realized. Figure 3.30 shows the structure of such a sensor, where a triple well is located to form three different PDs [280, 281]. This sensor has been commercialized as an APS type pixel.

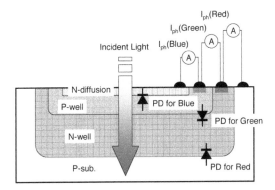

FIGURE 3.30: Device structure of image sensor with a triple junction. In the commercialized sensor, an APS structure is employed [281].

3.3.5.3 Controlling the potential profile

Changing the photo-sensitivity spectrum by controlling the potential profile has been proposed and demonstrated by many researchers [282–284]. The proposed systems mainly use a thin film transistor (TFT) layer consisting of multiple layers of p-i-i-i-n [283] and n-i-p-i-n [284]. Y. Maruyama *et al.* at Toyohashi Univ. Technology have proposed a smart CMOS image sensor using such a method [285, 286], although their aim was not color realization but filterless fluorescence detection, which is discussed in Sec. 3.4.6.2.

The principle of potential control is as follows [285, 286]. As discussed in Sec. 2.3.1.2, the sensitivity of a pn-junction PD is generally expressed by Eq. 2.19. Here we use the potential profile shown in Fig. 3.31. This figure is a variation of Fig. 2.10 wherein the NMOS-type PG is replaced with a PMOS-type PG on an n-type substrate, thus, giving two depletion regions, one originating from the PG and the other from the pn-junction. This pn-junction produces a convex potential that acts like a watershed for the photo-generated carriers.

In this case, the integral region in Eq. 2.18 is changed from 0 to x_c, where photo-generated carriers have an equal chance to flow to the surface or to the substrate. Carriers that flow to the substrate only contribute to the photocurrent.

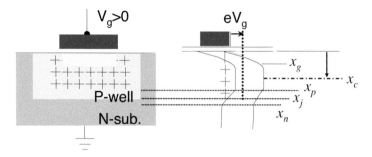

FIGURE 3.31: Device structure and potential profile for the filterless fluorescence image sensor.

The sensitivity thereby becomes

$$R_{ph} = \eta_Q \frac{e\lambda}{hc}$$
$$= \frac{e\lambda}{hc} \frac{\int_0^{x_c} \alpha(\lambda)P_o \exp\left[-\alpha(\lambda)x\right] dx}{\int_0^{\infty} \alpha(\lambda)P_o \exp\left[-\alpha(\lambda)x\right] dx} \tag{3.24}$$
$$= \frac{e\lambda}{hc}\left(1 - \exp\left[-\alpha(\lambda)x_c\right]\right).$$

From this, if two light beams with different wavelengths, excitation light *lambda*$_{ex}$ and fluorescence λ_{fl}, are incident simultaneously, then the total photocurrent I_{ph} is given by

$$I_{ph} = P_o(\lambda_{ex})A\frac{e\lambda_{ex}}{hc}\left(1 - \exp\left[-\alpha(\lambda_{ex})x_c\right]\right) + P_o(\lambda_{fl})A\frac{e\lambda_{fl}}{hc}\left(1 - \exp\left[-\alpha(\lambda_{fl})x_c\right]\right),$$
$$\tag{3.25}$$

where $P_o(\lambda)$ and A are the incident light power density with λ and the photo-gate (PG) area, respectively. When the photocurrent with two different gate voltages is measured, x_c has two values x_{c1} and x_{c2}, which results in two different photocurrents I_{ph1} and I_{ph2}:

$$I_{ph1} = P_o(\lambda_{ex})A\frac{e\lambda_{ex}}{hc}\left(1 - \exp\left[-\alpha(\lambda_{ex})x_{c1}\right]\right) + P_o(\lambda_{fl})A\frac{e\lambda_{fl}}{hc}\left(1 - \exp\left[-\alpha(\lambda_{fl})x_{c1}\right]\right),$$
$$I_{ph2} = P_o(\lambda_{ex})A\frac{e\lambda_{ex}}{hc}\left(1 - \exp\left[-\alpha(\lambda_{ex})x_{c2}\right]\right) + P_o(\lambda_{fl})A\frac{e\lambda_{fl}}{hc}\left(1 - \exp\left[-\alpha(\lambda_{fl})x_{c2}\right]\right).$$
$$\tag{3.26}$$

In these two equations, the unknown parameters are the input light intensities $P_o(\lambda_{ex})$ and $P_o(\lambda_{fl})$. We can calculate the two input light powers, $P_o(\lambda_{ex})$ for the excitation light and $P_o(\lambda_{fl})$ for the fluorescence power, that is, filterless measurement can be achieved.

3.3.5.4 Sub-wavelength structure

The fourth method to realize color detection is to use a sub-wavelength structure such as a metal wire grid or surface plasmons [287–291] and photonic crystals [167,292]. These technologies are in their preliminary stages but may be effective for CMOS image sensors with fine pitch pixels. In sub-wavelength structures, the quantum efficiency is very sensitive to polarization as well as the wavelength of the incident light and the shape and material of the metal wire grid. This means that the light must be treated as an electro-magnetic wave to estimate the quantum efficiency.

When the diameter of an aperture d is much smaller than the wavelength of the incident light λ, the optical transmission through the aperture T/f, which is the transmitted light intensity T normalized to the intensity of the incident light in the area of the aperture f, decreases according to $(d/\lambda)^4$ [293], which causes the sensitivity of an image sensor to decrease exponentially. T. Thio *et al.* reported a transmission enhancement through a sub-wavelength aperture surrounded by periodic grooves on a metal surface [294]. In such a structure, surface plasmon (SP) modes are excited by the grating coupling of the incident light [295], and the resonant oscillation of the SPs causes an enhancement of the optical transmission through the aperture. This transmission enhancement could make it possible to realize an image sensor with a sub-wavelength aperture. From the computer simulation results given in Ref. [296], an aluminum metal wire grid enhances the optical transmission, while a tungsten metal wire grid does not enhance it. The thickness and the line and space of the metal wire grid also influence the transmission.

The other application of SP modes is in color filters [290, 297]. Figure 3.32(a) shows color filters in concentrically periodic corrugated structure or bullseye structure made of a metal. A hole is fabricated in the center of the structure. The wavelength of the light can be selected by changing the corrugation period on metal surface as shown in Fig. 3.32(b).

Since the surface plasmon filter is made of metal, the transmittance is essentially degraded owing to the opaque nature of the metal. To alleviate this issue, deflectors or splitters of nano-structure made of dielectric materials have been developed [298–300]. Since the nano-deflectors can be made of dielectric material such as SiN, the structure is transparent and hence, the transmittance is barely degraded. In addition, the materials, such as SiN are conventionally used in standard CMOS process so that the nano-deflectors are compatible with it. One issue here is that all of the input light cannot be deflected, i.e., a part of the input light is not deflected. To manage this issue, two nano-deflectors are used in combination with the image processing.

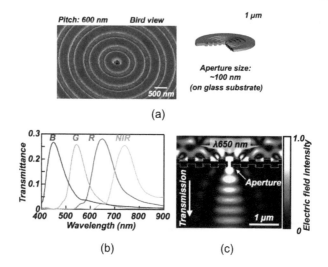

(a)

(b)   (c)

FIGURE 3.32: Metallic color filter with concentric periodical corrugation: (a) SEM images of the silver thin film with a concentrically periodic corrugation; (b) simulation results of transmission spectra for blue, green, red and near-infrared; (c) simulation results of electric field intensity distribution at the incident wavelength of 650 nm [290]. By courtesy of Prof. Ono at Sizuoka Univ.

3.4   Dedicated pixel arrangement and optics for smart CMOS image sensors

This section describes smart CMOS image sensors that have dedicated pixel arrangement and dedicated optics. A conventional CMOS image sensor uses a lens to focus an image onto the image plane of the sensor, where pixels are placed in an orthogonal configuration. In some visual systems, however, non-orthogonal pixel placement is used. A good example is our vision system where the distribution of the photo-receptors is not uniform. Around the center or fovea, they are densely placed, while in the periphery they are sparsely distributed [199]. This configuration is preferable as it is able to detect an object quickly with a wide angle view. Once the object is located, it can be imaged more precisely by the fovea by making the eye move to face the object. Another example is an insect with compound eyes [301, 302]. A special placement of pixels is sometimes combined with special optics, such as in the compound eyes of insects. Pixel placement is more flexible in a CMOS image sensor than in a CCD sensor, because the alignment of the CCDs is critical for charge transfer efficiency. For example, a curved placement of a CCD may degrade the charge transfer efficiency.

In this section, some special pixel arrangements for smart CMOS image sensors are described. Subsequently, some smart CMOS sensors with dedicated optics are introduced.

3.4.1 Phase-difference detection auto focus

CMOS image sensors with fast auto focus function can be realized to employ phase-difference detection in the image plane [303, 304]. In the sensor, some pixels are used for PDAF (phase-difference detection auto focus). Part of the PD is light shielded so that the pixels for PDAF cannot be used for imaging. To alleviate this issue, a split micro-lens is fabricated on a pixel. The split micro-lens can produce angular dependence so that it can be used for PDAF, while the addition of split signals can be used for imaging. This method is effective for CMOS image sensors with a small pitch [305].

3.4.2 Hyper omni vision

Hyper omni vision (HOVI) is an imaging system that can capture a surrounding image in all directions by using a hyperbolic mirror and a conventional CCD camera [306, 307]. The system is suitable for surveillance. The output image is projected by a mirror and thus is distorted. Usually, the distorted image is transformed to an image rearranged with Cartesian coordinates and is then displayed. Such an off-camera transformation operation restricts the available applications. A CMOS image sensor is versatile with regard to pixel placement. Thus, pixels can be configured so as to adapt for a distorted image directly reflected by a hyperbolic mirror. This realizes an instantaneous image output without any software transformation procedure, which opens up various applications. In this section, the structure of a smart CMOS image sensor for HOVI and the characteristics of the sensor are described [308]. A conventional HOVI system consists of a hyperbolic mirror, lens, and CCD camera. Images taken by HOVI are distorted owing to the hyperbolic mirror. To obtain a recognizable image, a transformation procedure is required. Usually this is accomplished by software in a computer. Figure 3.33 illustrates the imaging principle of HOVI.

An object located at $P(X,Y,Z)$ is projected to a point $p(x,y)$ in the 2D image plane by a hyperbolic mirror. The coordinates of $p(x,y)$ are expressed as follows.

$$x = \frac{Xf\left(b^2 - c^2\right)}{\left(b^2 + c^2\right)Z - 2bd\sqrt{X^2 + Y^2 + Z^2}}, \tag{3.27}$$

$$x = \frac{Yf\left(b^2 - c^2\right)}{\left(b^2 + c^2\right)Z - 2bd\sqrt{X^2 + Y^2 + Z^2}}. \tag{3.28}$$

Here, b and c are parameters of the hyperbolic mirror and f is the focal length of the camera.

A smart CMOS image sensor is designed to arrange pixels in a radial pattern according to Eqs. 3.27 and 3.28 [308]. A 3T-APS is used for the pixel circuits. A

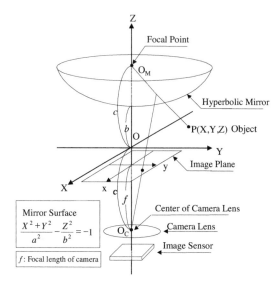

FIGURE 3.33: Configuration of a HOVI system.

feature of the chip is that the pitch of the pixel becomes smaller from the outside edge to the center. Thus, four types of pixels with different sizes are employed. For the radial configuration, vertical and horizontal scanners are placed along the radial and circular directions, respectively. A microphotograph of the fabricated chip is shown in Fig. 3.34. In the figure, a close-up view around the inner-most pixels is also shown.

Figure 3.34(c) illustrates the image acquisition for the conventional HOVI with a CCD camera. The output image is distorted. In Fig. 3.34(d) , the image was captured by the fabricated image sensor. An image without distortion was obtained.

3.4.3   Biologically inspired imagers

Some animals have eyes with different structures from conventional image sensors. These specific structures have some advantages such as a wide field-of-view (FOV). This section shows some examples of such biologically inspired image sensors.

3.4.3.1   Curved image sensor

Curved image sensor is inspired by some animal eyes which have a curved detection surface. Mimicking them, curved image sensors have been published [309–311] as shown in Fig. 3.35. Curved image sensors can achieve a wide FOV and low aberrations with a simple lens system. In Ref. [310], a curved BSI CMOS image sensor was fabricated. By curving the sensor, the F number can be reduced and thus the system sensitivity can be increased. In addition, when the sensor is curved, the sensor has a tensile stress. This tensile stress reduces the bandgap of Si so that the

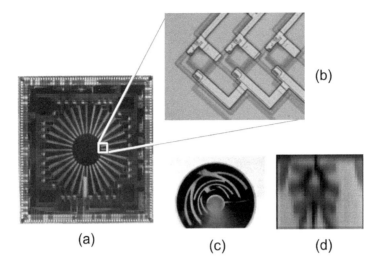

FIGURE 3.34: Microphotograph of a smart CMOS image sensor for a HOVI system (a) and its close-up microphotograph (b). Input of a Kanji character taken by a conventional HOVI system (c) and its output images for the sensor (d).

dark current is reduced. This side effect is preferable for the image sensor so as to improve its performance.

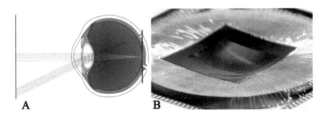

FIGURE 3.35: Curved image sensor. Adapted from [311] with permission.

3.4.3.2 Compound eye

A compound eye is a biological visual system in arthropods including insects and crustaceans. There are a number of independent tiny optical systems with small FOV as shown in Fig. 3.36. The images taken by each of the independent tiny eyes, called *ommatidium*, are composited in the brain to reproduce a whole image.

The advantages of a compound eye are its wide FOV with a compact volume and a short working distance, which can be realized with an ultra-thin camera system. Furthermore, only a simple imaging optics is required for each *ommatidium*, because only a small FOV is required for an *ommatidium*. The disadvantage is relatively poor

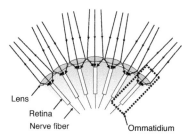

FIGURE 3.36: Concept of an apposition compound eye system. The system consists of a number of *ommatidum*, which are composed of a lens, retina, and nerve fiber. It is noted that another kind of compound eye is the neural superposition eye.

resolution.

Artificial compound eyes have been developed by many organizations [312–318]. In the following sections, one example of smart image sensor systems using the compound eye architecture, TOMBO developed by J. Tanida *et al.* of Osaka University is introduced.

TOMBO The TOMBO* system, an acronym for thin observation module by bound optics, is another compound eye system [319, 320]. Figure 3.37 shows the concept of the TOMBO system. The heart of the TOMBO system is a number of optical imaging systems, each of which consists of several micro-lenses, called imaging optical units. Each imaging optical unit captures a small but full image with a different imaging angle. Consequently, a number of small images with different imaging angles are obtained. A whole image can be reconstructed from the compound images from the imaging optical units. A digital post-processing algorithm enhances the composite image quality.

A crucial issue in realizing a compound eye system is the structure of the ommatidium with the micro-optics technology. In the TOMBO system, the signal separator shown in Fig. 3.37 resolves this issue. A CMOS image sensor dedicated for the TOMBO system has been developed [119, 321]. The TOMBO system can also be used as a wide angle camera system as well as a thin or compact camera system.

3.4.4   Light field camera

A light field camera is called as a plenoptic camera. The concept of light field is shown in Fig. 3.38 [322]. The light is associated with two planes u–v and x–y, and thus, four parameters (u,v,x,y) as shown in Fig. 3.38 (a) . A conventional image sensor only detects the two-dimensional plane, i.e., (x,y).

---

*TOMBO is the Japanese word for "dragonfly". A dragonfly has a compound eye system.

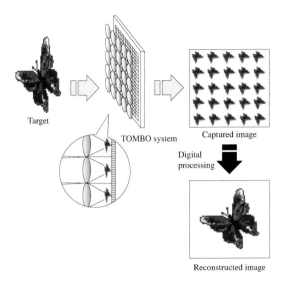

FIGURE 3.37: Concept of TOMBO. Courtesy of Prof. J. Tanida at Osaka Univ.

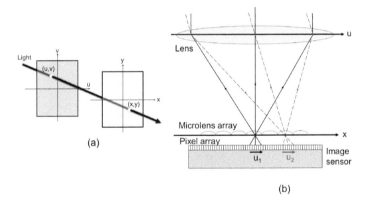

FIGURE 3.38: Concept of light field image sensor. (a) The light is associated with four parameters (u,v,x,y). (b) Schematic structure of the light field image sensor.

The light field image sensors have been developed to detect the other two parameters, (u, v). The concept of the light field image sensors is shown in Fig. 3.38 (b). The imaging lens is focused on the x–y surface. Since Fig. 3.38 (b) is drawn as a one-dimensional figure, only the x axis is shown in the figure. In the figure, two rays are drawn, one of which is a solid line and the other a dashed line according to the incident angles with the lens. In the focal plane, an array of micro-lens is placed. The exit rays from the micro-lenses are projected on the image sensor surface, where the coordinate is u on the u–v plane. This projected image plane or image sensor plane is the lens plane image, but each micro-lens corresponds to the direction of the ray. Since all of the rays are recorded in this sensor, it is possible to reconstruct the focus plane, and by recording the direction of the ray, a 3D image can be reconstructed [322], which is a 3D range finder in Sec. 4.6 of Chapter 4.

It is difficult to align the micro-lens array in an image sensor with a small pixel pitch. Wang *el al.* have demonstrated the light field image sensors by using two sets of diffraction grating array on the surface of an image sensor [323–325] as shown in Fig. 3.39. The diffraction grating is fabricated by using metal wires in interconnected metal layers in a standard CMOS process, which is similar to that in integrated polarization image sensors described in Sec. 3.4.5. In this sensor, two grating layers are embedded in a sensor and can detect incident angle of the input light. This property can achieve a 3D image. The pixel is called an ASP (angle sensitive pixel) [323]. The input light is diffracted by the top metal grating and produces an interference pattern in the depth of the second grating. The second grating passes or blocks the interference pattern. Since the position of the interference pattern is dependent on the incident angle of the input light, the pixel acts as ASP.

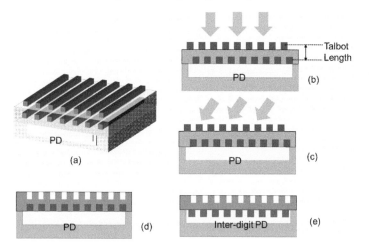

FIGURE 3.39: Schematic structure of the light field image sensor based on Talbot effect. Illustrated after [323].

By using diffractive optics, another type of light field image sensor has been reported [326]. In this case, the right and left beams can be separated by a lenticular lens as a beam splitter and digital micro-lens acts as focus enhancer. The digital micro-lens is based on a diffractive optics. The schematic structure is described in Sec. 4.6.3.1 of Chapter 4.

3.4.5 Polarimetric imaging

Polarization is a characteristic of light [302]. Polarimetric imaging makes use of polarization and is used in some applications to detect objects more clearly, for example, the reflected image from glass can be clearly taken by using polarimetric imaging. In addition, the polarimetric imaging is used to measure 2D birefringent distribution of transparent materials, which shows the residual stress inside the materials. Polarimetric imaging is described in details in the book by Sarkar and Theuwissen [327]. Some types of polarimetric image sensors have been firstly developed by placing birefringent materials such as polyvinyl alcohol (PVA) [328, 329], Rutile (TiO_2) crystal [330], YVO_4 crystal [331], liquid crystal polymer (LCP) [332], and photonic crystal [333] on the sensor surface. The other type of polarizer is a metal wire grid as shown in Fig. 3.40. When linearly polarized light hits on a metal gwire rid the pitch of which is much smaller than the wavelength of the light, the intensity of the light parallel to the grid is attenuated because it excites oscillated current in the grid, some of which is converted into heat and the remaining produces electric field with anti-phase against the input electric field. On the contrary, the linearly polarized light perpendicular to the grid array is hardly attenuated. Wire grid polarizers are widely used in NIR region because fine grid pitch is not required for the NIR light. Birefringent materials show excellent polarimetric property in visible region; it is, however, difficult to integrate the material with an image sensor.

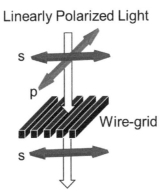

FIGURE 3.40: Conceptual illustration of metal wire grid polarizer.

In 2008, NAIST (Nara Institute of Science and Technology) group has proposed to use a metal wire in CMOS process as wire grid and successfully demonstrated its use in analyzing polarization [334] as shown in Fig. 3.41. The other groups have also demonstrated the same structure [335–337].

This structure is suitable for the CMOS fabrication process because metal layer for wiring can be used as the wire grid polarizer. Fine fabrication process such as 65 nm CMOS process is more suitable for the wire grid polarizer, and thus has achieved higher extinction ratio (ER) compared with the 0.35 μm or 0.18 μm CMOS process

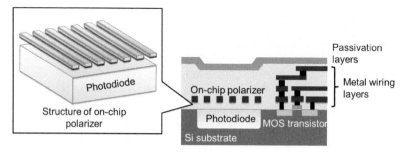

FIGURE 3.41: Concept of polarization detection image sensor.

[338]. Here ER is defined as the ratio of the maximum to minimum normalized intensity in the angular profile, which is a parameter of polarizer performance. Figure 3.42 shows experimental results of captured images that have local polarization variation by using a polarimetric image sensor in 65 nm CMOS process [339]. Table 3.3 summarizes the specifications of integrated polarization CMOS image sensors including the CCD-based sensors.

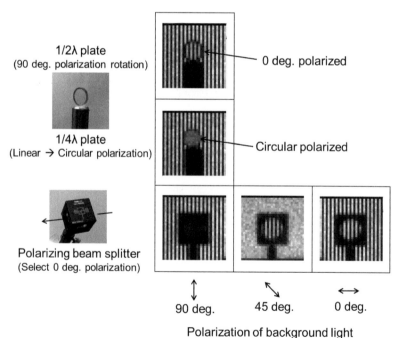

FIGURE 3.42: Captured images that have local polarization variation. Adapted from [339] with permission.

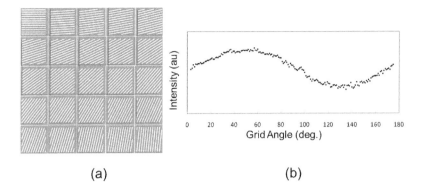

(a)   (b)

FIGURE 3.43: Polarization detection image sensor. (a) Grid embedded pixel array. (b) The pixel output as a function of grid angle.

The polarimetric image sensors with integrated wire grid arrays in 0.35 μm standard CMOS technology has low extinction ratio because the wire grid pitch is comparable with the visible wavelength. To leverage this low extinction ratio, micro-polarizers with a grid orientation from 0 to 180° in each degree and this set is repeated so as to average the fluctuations. Figure 3.43 shows the pixel array embedded with a micro-polarizer, the orientation of which is varied from 0 to 180°. In Sec. 5.3.1 of Chapter 5, this type of polarimetric CMOS image sensor is demonstrated to be used in chemical applications, where polarization is frequently used to identify chemical substances.

The integrated CMOS polarization image sensors embedded the wire grid on the chip was first commercialized [337,340]. They can be used for industrial applications such as inspection of scratch and dirt on glass. Wire grid type polarizer is integrated on a sensor chip. Each wire is separated with an air gap instead of an insulator to ensure the wavelength of incident light is not changed inside the wire grid. In addition, the wire grid is made of aluminum instead of copper which is usually used as a metal wiring layer, due to its lower absorption than that of copper.

TABLE 3.3
Comparison of specifications of polarimetric image sensors

Affiliation	Year	Sensor	Orientation (°)	Grid L/S (nm)	Material	Fab. method	ER (dB)	Ref.
Johns Hopkins Univ.	1996	2-μm standard CMOS (PTr)	0 & 90	–	Patterned PVA	Separate	–	[328]
	1998	1.2-μm standard CMOS (PTr)	0 & 90	–	Patterned birefringent crystal	Separate	–	[330]
State Univ. NY	2006	1.5-μm standard CMOS (current mode)	0 & 90	–	Patterned birefringent crystal	Separate	–	[331]
Tohoku Univ. & Photonic Lattice	2007	CCD	0, 45, 90 & 135	–	Dielectric photonic crystal	Separate	40	[333]
NAIST	2008	0.35-μm standard CMOS (3T-APS)	0–180 (1° step)	600/600	CMOS metal wire	CMOS process	3	[334]
	2013	65-nm standard CMOS (3T-APS)	0 & 90	100/100	CMOS metal wire	CMOS process	19	[338]
Univ. Pennsylvania	2008	0.18-μm standard CMOS (modified 3T-APS)	0 & 45	–	Patterned PVA	Separate	17	[329]
	2014	65-nm standard CMOS (3T-APS)	0, 45 & 90	90/90	CMOS metal wire	CMOS process	17	[341]
TU Delft & IMEC	2009	0.18-μm CIS CMOS (4T-APS)	0, 45 & 90	240/240	CMOS metal wire	CMOS process	8.8	[342]
Univ. Washington	2010	CCD	0, 45, 90 & 135	70/70	Al nanowire	Separate	60	[343]
& Univ. Illinois	2014	0.18-μm CIS CMOS (4T-APS)	0, 45, 90 & 135	70/70	Al nanowire	Separate	60	[344]
Univ. Arizona	2012	CCD	0, 45, 90 & 135	–	Patterned LCP	Separate	14	[332]
SONY	2018	90-nm BSI CIS CMOS (4T-APS)	0, 45, 90 & 135	50/100	Air-gap wire grid	On-chip post process	85	[340]

3.4.6   Lensless imaging

This section describes lensless imaging or lens-free imaging. Lensless imaging is an imaging system without any lens. The conventional lens is refractive, while the Fresnel lens is diffractive. Here, an imaging system with a diffractive lens is not included in lensless imaging. Lensless imaging systems are divided into two types as shown in Fig. 3.44 [345].

3.4.6.1   Coded aperture camera

The lensless imaging system, in which a mask is placed on or close to the sensor surface to modulate the input image (Fig. 3.44(b)) is called a coded aperture lensless imaging system [346–352]. This mask acts as a multiple pinhole array [347]. A pinhole camera is the first lensless camera, but has a disadvantage of low efficiency of usage of light. To overcome it, coded aperture methods have been developed.

The coded aperture cameras were initially used for X-ray and gamma ray imaging, since appropriate lens materials did not exist in these wavelength regions [347, 348]. Coded apertures spatially encode the image, and the retrieval process is as follows [347]. The recorded image R is the convolution between the object O and the aperture A, or $R = O \times A$. The decoding process G recovers the original image as $R \otimes G = \hat{O}$. Consequently, the recovered original image $\hat{O}$ is expressed as,

$$\hat{O} = R \otimes G = (O \times A) \otimes G = O * (A \otimes G). \tag{3.29}$$

Thus,

$$\text{if } A \otimes G = \delta \text{ (delta function), then } \hat{O} = O. \tag{3.30}$$

To retrieve the original image to the best possible extent with minimum computational load, the aperture design is critical to satisfy Eq. 3.30. Several aperture designs have been developed such as uniformly redundant array [346], pseudo-noise pattern [351], planar Fourier captured array (PFCA) [349], spiral pattern [350], Fresnel zone [352], etc.

By taking advantage of lensless cameras, compact cameras have been reported such as "FlatCam" [351], "PicoCam" [350], and others [349]. In FlatCam, the coded mask is placed close to the sensor surface (at distance of 0.5mm) so that thin form factor can be achieved [351]. In PicoCam, the code aperture is a phase grating made of glass so that the light loss is reduced compared to the amplitude coded aperture. The integrated coded aperture in a CMOS image sensor has been reported in [349]. In this case, a chip-size camera can be realized. The coded aperture pattern is fabricated with the metal layers used in a conventional CMOS fabrication process. The structure is the same as the one in the light field sensor based on Talbot effect in Fig. 3.39, so that the grating pattern modulates the input image with sinusoidal intensity. To realize the coded aperture array, the grating pattern in each pixel is placed with different slant angle like in Fig. 3.43(a), as well as a set of different pitch of the grating. Each pixel modulates the intensity of the input image sinusoidally with the incident angle. All the pixels can achieve a complete 2D Fourier transform. The main drawback of these compact cameras with coded apertures is a lower resolution compared with the conventional cameras with lens. To increase the pixel number, this drawback is alleviated to some extent.

The pattern is spatially fixed in coded aperture cameras but some types of coded apertures introduced programmable apertures [353–355]. SLMs (spatial light modulators) are used as programmable coded apertures, where transparency can be controlled in each pixel independently. By introducing programmable coded apertures, instantaneous field of view change, split field of view and optical computation during image formation can be achieved [353]. As an example of a camera being used as a computational sensor, an optical correlation processing can be executed to display a correlation template image on the programmable coded apertures.

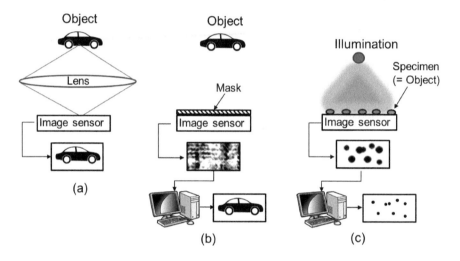

FIGURE 3.44: Concept of lensless imaging: (a) conventional imaging system with lens; (b) lensless imaging system with mask; (c) lensless imaging system with direct contact of specimen on the sensor surface.

3.4.6.2 Direct contact type

In the second type of lensless imaging system, direct contact or close contact with the specimen is ensured to produce the image [356–358]. The main application for this type is in the observation of cells, for example, in an optical microscope [357] and to detect specific cells in microfluidic devices or flow cytometers [359, 360]. Without a lens, a large FOV with a compact size can be achieved. This type is further classified into diffraction-based imaging (or shadow imaging) and contact-mode imaging methods. The diffraction-based imaging method was developed by Ozcan *et al.* by using the holographic method to enhance the spatial resolution [357]. When light hits on a cell from above, the scattered light and the straight light are interfered with and they produce an interference pattern or the holographic pattern.

To analyze this pattern, spatial resolution can be enhanced better compared with the direct projection type. Optical microscopes with lensless image sensors are compact with large (FOV. Furthermore, it can be applied to flow cytometer, which is used for detecting specific types of cells in flowing cells in a microfluidic device. In a conventional flow cytometry, a camera is placed to detect specific cells. By introducing the lensless imaging system, the total volume size of the system can be reduced and thus can be applied at the point-of-care (POC).

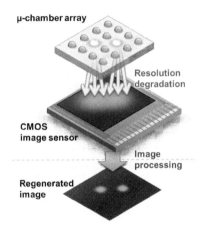

FIGURE 3.45: Lensless imaging system based on the direct contact method. A fiber optic plate is used in this case to transfer the image with minimal image degradation. Adapted from [361] with permission.

The other method for contact mode imaging is to use an image wherein a specimen is directly placed on a sensor surface [362, 363]. This type is mainly used for fluorescent imaging, because fluorescence emits in every direction with a wavelength different from the wavelength of the excitation light, so that no interference is produced and thus, the holographic methods are difficult to apply. In this case, the degradation of resolution is more critical, because more space between the specimen and the sensor surface is produced by inserting an emission filter on the sensor surface to eliminate the excitation light. Figure 3.45 shows the contact type for fluorescence imaging [361]. To suppress the excitation light intensity compared to the fluorescence light intensity, several methods have been reported such as using a prism [364], nano-plasmonic filter [365], interference filter [363], light pipe structure [366], combination of absorption and interference filters [362, 367], etc. It is not difficult to suppress the excitation light if the Stokes shift is large. The Stokes shift means the difference between the peak wavelength of excitation light and the fluorescence light. For example, the difference of the peak wavelength between the excitation and fluorescence light in Hoechst®3342 is about 120 nm, while in GFP, it is about 30 nm. Thus, in this case, detecting the GFP is more difficult than in Hoechst®3342.

4

Smart imaging

4.1 Introduction

Some applications require imaging that is difficult to achieve by using conventional image sensors, either because of the limitations in their fundamental characteristics, such as speed and dynamic range, or because of their need for advanced functions, such as distance measurement. For example, intelligent transportation system (ITS) and advanced driver assistance system (ADAS), in the near feature, will require intelligent camera systems to perform lane keeping assist, distance measurement, driver monitoring, etc. [368], for which smart image sensors must be applied with a wide dynamic range of over 100 dB, higher speeds than the video rate, and the ability to measure distances of multiple objects in an image [369]. Security, surveillance, and robot vision are applications that are similar to ADAS. Smart imaging is also effective in information and communication fields, as well as in biomedical engineering.

Numerous implementation strategies have been developed to integrate the smart functions described in the previous chapter. These functions are commonly classified by the level at which the function is processed, namely pixel level, column level, and chip level. Figure 4.1 shows smart imaging in CMOS image sensors, illustrating this classification. Of course, a combination of these levels in one system is also possible.

The most straightforward implementation is the chip-level processing, wherein the signal processing circuits are placed after the signal output, as shown in Fig. 4.1 (a). This is an example of a 'camera-on-a-chip', where an ADC, a noise reduction system, a color signal processing block, and other elements are integrated on a chip. It should be noted that this type of processing requires almost the same output data rate; therefore, the gain in the signal processing circuits is limited. The second implementation method is the column-level processing or column-parallel processing. This is suitable for CMOS image sensors, because the column output lines are electrically independent. As signal processing is achieved in each column, a slower processing speed can be used than for chip-level processing. Another advantage with this implementation is that the pixel architecture can be the same as in a conventional CMOS image sensor. Hence, for example, a 4T-APS can be used. This feature is a great advantage in achieving a good SNR .

The third implementation method is pixel-level processing or pixel-parallel

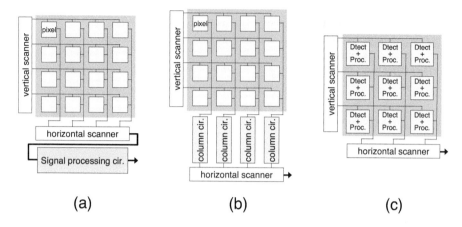

FIGURE 4.1: Basic concept of smart CMOS image sensors: (a) Chip-level processing; (b) column-level processing; and (c) pixel-level processing.

processing. In this method, each pixel has a signal processing circuit as well as a photodetector. This can realize fast, versatile signal processing, although the photodetection area or fill factor is reduced so that the image quality may be degraded compared with the former two implementation methods. Also, in this method, it is difficult to employ a 4T-APS for a pixel. However, this architecture is attractive and is a candidate for the next generation of smart CMOS image sensors. Especially, recently developed stacked technology can open a new path for pixel-level processing.

In this chapter, we survey smart imaging required for smart CMOS image sensors for several applications.

4.2 High sensitivity

High sensitivity is essential for imaging under low light in several scientific applications in fields, such as astronomy, biotechnology, and also in consumer and industrial applications in industries such as automobile and surveillance. Some image sensors have ultra-high sensitivity, for example, super HARP [100] and EMCCD (Electron Multiplying CCD) [370]. However, in this section we focus on smart CMOS image sensors with high sensitivity.

4.2.1 Dark current reduction

Some applications in low light imaging do not require video rate imaging and thus, long accumulation times are allowed. For long exposure times, dark current and flicker noise or $1/f$ noise are dominant. A detailed analysis for suppressing noise in low light imaging using CMOS image sensors is reported in Refs. [210,211,371]. To decrease the dark current of a PD, cooling is the most effective and straightforward method. However, in many applications it is difficult to cool the detector.

Here, we discuss how to decrease the dark current at room temperature. First of all, pinned PDs (PPDs) or buried PDs (BPDs), as described in Sec. 2.4.3, are effective in decreasing the dark current.

Decreasing the bias voltage at the PD is also effective [80]. As mentioned in Sec. 2.4.3, the tunnel current strongly depends on the bias voltage. Figure 4.2 shows near zero bias circuits, developed as reported in Ref. [372]. Here, one near zero bias circuit is located in a chip and provides the gate voltage of the reset transistor.

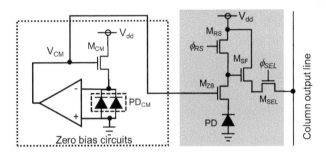

FIGURE 4.2: Near zero bias circuits for reducing dark current in an APS. Illustrated after [372].

4.2.1.1 PFM for low light imaging

The PFM photo-sensor can achieve low light detection with near-zero bias of the PD and can obtain a minimum detectable signal of 0.15 fA (1510 s integration time), as shown in Fig. 4.3 [210]. As mentioned in Sec. 2.3.1.3, the dark current of a PD is exponentially dependent on the PD bias voltage. Therefore, the PD bias near zero voltage is effective in reducing the dark current. In such a case, special care must be taken to reduce the other leakage currents.

In a PFM photo-sensor, with a constant PD bias, the leak current or sub-threshold current of the reset transistor M_{rst}, as depicted in Fig. 4.3 is critical and must be reduced as much as possible. To address the issue, Bolton *et al.* have introduced the reset circuit shown in Fig. 4.3, wherein the original reset transistor M_{rst} is replaced by a T-switch circuit. The T-switch circuit consists of two transistors, M_{rst1} and M_{rst2}, and the connection node voltage of the two transistors V_D is near zero when they are connected to the ground through the turn-on transistor. Consequently, the drain–source voltage of the reset transistor M_{rst1} is near zero because $V_{PD} \approx 0$ and hence, the sub-threshold current reaches zero [211]. It should be noted that the sub-threshold current is exponentially dependent on the drain—source voltage as well as the gate–source voltage, as discussed in Appendix F.

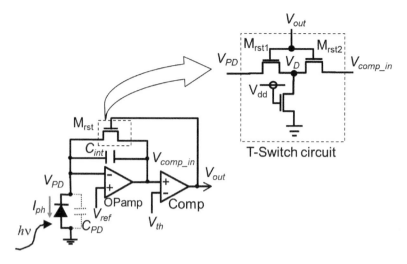

FIGURE 4.3: PFM circuits with constant PD for low light detection. The single reset transistor M_{rst} is replaced by a T-Switch circuit [211]. OP amp: operational amplifier, Comp: comparator. Illustrated after [210].

4.2.2 Differential APS

The differential APS has been developed to suppress the common mode noise, as reported in Ref. [371], and the same is shown in Fig. 4.4. Here, the sensor uses a pinned PD and low bias operation using a PMOS source follower. It should be noted that the amount of $1/f$ noise in a PMOS is less than that in an NMOS. Consequently, the sensor demonstrates ultra-low light detection of 10^{-6} lux over 30 s of integration time at room temperature.

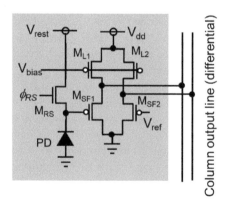

FIGURE 4.4: Differential APS. Illustrated after [371].

4.2.3 High conversion gain pixel

The conversion gain, g_c is one of the critical parameters for high sensitivity, and is defined in Sec. 2.5.3 as C_{FD}^{-1}, where C_{FD} is the capacitance of the FD. Thus, high sensitivity can be achieved to reduce the capacitance of FD. The capacitance of FD consists of the junction capacitance of FD and other parasitic capacitance elements. Consequently, even if the junction capacitance of the FD is reduced, the other parasitic capacitance elements still retain their values. In Ref. [373], a reset gate transistor is eliminated so that all the capacitances associated with the reset transistor are diminished. This pixel structure without a reset transistor, called RGL (reset gate less) has a drain terminal that is placed close to the FD. By resetting the FD, the drain is biased so deeply that all of charges in the FD can be flowed into the drain. The image sensor with RGL pixel was achieved with $g_c = 220\mu V/e^-$, and a sub-electron read noise of $0.27e_{rms}^-$.

4.2.4 SPAD

To achieve ultra-low light detection with a fast response, the use of an APD is beneficial. It should be noted that the APD works in analog mode and thus, it is difficult to control its gain in an array device. On the contrary, in the Geiger mode, the APD (in this case the photodiode is called a single photon avalanche diode (SPAD)) produces a spike-like signal, not an analog output, and also VAPD produces digital signals described in Sec. 2.3.4.2, both of which are suitable for an array device. A pulse shaper with an inverter is used to convert the signal to a digital pulse, as shown in Fig. 3.15. The details have been described in Sec. 3.2.2.3 of Chapter 3.

4.2.5 Column-parallel processing

The output of each pixel in a CMOS image sensor connects a column output line and hence, column-parallel processing is possible. A CDS, an ADC, and an amplifier are placed in each column. By modifying these circuits, it is possible to improve the sensitivity. In the following sections, three circuits, namely correlated multi-sampling, adaptive gain control, and ADC with digital CDS, are introduced.

Correlated Multi-Sampling Conventional CDS is used for sampling reset and signal level as described in Sec. 2.6.2.2 of Chapter 2. Correlated multi-sampling (CMS) likewise for sampling multiple resets and multiple signals as shown in Fig. 4.5 [373, 374]. Similar multi-sampling techniques to achieve high sensitivity have been reported in [375, 376]. Such multi-processing methods basically increase the SNR by $\sqrt{M}$ times, where M is the processing number, although other noise factors affect the SNR value as well. In Fig. 4.5 (a), the column circuits consist of sampling and integration parts, and ADC. The reset is sampled and integrated, and this process is repeated by M times. Then, the final value is stored in C_{SH}, converted by the ADC, and stored in the memory for reset. The signal sampling follows the same process. The drawback of this method is that it takes longer to achieve high sensitivity.

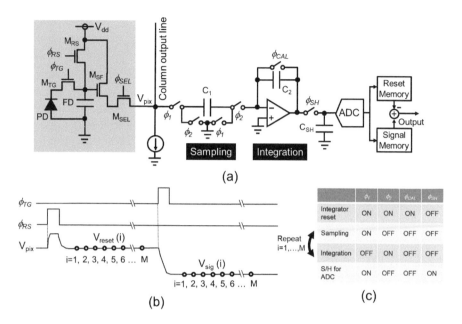

(a)

(b)

(c)

FIGURE 4.5: Correlated Multi-Sampling. Circuits: (a), circuits, (b) timing chart, and (c) switching status. In [374], RGL pixel structure was used instead of a conventional 4T-APS. Illustrated after [374].

Adaptive gain control   As the gain in a column increases, the input-referred noise decreases and, therefore, the SNR increases. However, a high gain saturates the signal and as a result, the dynamic range decreases. On the contrary, such a high gain is not required for a pixel with a high signal level. Thus, introducing an adaptive gain control to each pixel value can solve these issues [377–379].

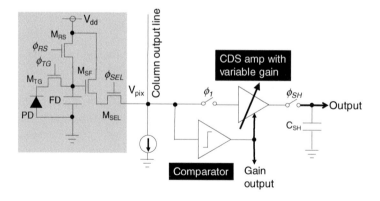

FIGURE 4.6: Adaptive gain control. Illustrated after [379].

Figure 4.6 shows an example of an adaptive gain control. The gain value is selected to one or eight according to the comparison result by the comparator in Fig. 4.6. The programmable amplifier is modified with the CDS circuits as shown in Fig. 2.31 in Sec. 2.6.2.2 and it has both functions of both amplification and CDS . In this case, the capacitor C_2 shown in Fig. 2.31 is changed to two capacitors, C_{21} and C_{22}. The value of the input capacitance C_1 in Fig. 2.31, in this case, is $8 \times C_{21}$. C_{22} is equal to $7 \times C_{21}$. By this configuration, the gain value can be changed by one and eight. If the signal level is smaller than a certain value, the gain is set to eight, otherwise set to one.

ADC with digital CDS   It is effective to merge the functions of ADC and CDS in the column circuits. Figure 4.7 shows such an example [380, 381].

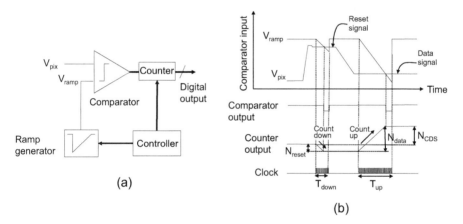

(a)

(b)

FIGURE 4.7: Sequence of CDS operation by using SS-ADC. An up and down counter is used in this case instead of a simple counter as shown in Fig. 2.33; (a) Schematic diagram, (b) Timing chart. Illustrated after [380].

The column circuitry as shown in Fig. 4.7 (a) is basically a SS-ADC, which is described in Sec. 2.6.3. As shown in the timing chart of Fig. 4.7 (b), in the reset sampling duration, the counter in the SS-ADC outputs the counting number downward. On the contrary, in the data sampling duration, the counter outputs the counting number upward. Thus, the final counting number is the difference between the number of up-counting and the number of down-counting, and is equal to the digital value of the difference between the data and reset signals. Eventually, the CDS operation is completed in the digital domain. This architecture is effective not only for high sensitivity but also for high speed operation, because CDS and ADC operations are simultaneously performed in column processing, and thus, are less affected by noise even in a high speed operation.

4.3   High-speed

4.3.1   Overview

This section presents high speed CMOS image sensors. They are classified into three categories. The first category is the CMOS image sensors with a global shutter. As described in Sec. 2.10 of Chapter 2, the conventional CMOS image sensors work in the rolling shutter mode, while CCDs work in the global shutter mode. The global shutter function can correctly acquire objects that move with high-speed over its frame rate. Thus, this function is preferable for consumer cameras such as automobile cameras.

The other category is the CMOS image sensors with a high frame rate (i.e., over 1 kfps). A high frame rate alleviates the disadvantage of a rolling shutter that is inherent to CMOS image sensors. It should be noted that the frame rate is not a good criterion for high speed in some cases, because it depends on the pixel number. Thus the pixel rate or data rate is a better criterion, especially in CMOS image sensors with high resolution. The pixel rate R_{pix} is defined as the output pixel counts per second, and expressed as

$$R_{pix} = N_{row} \times N_{col} \times R_{frame}, \tag{4.1}$$

where N_{row}, N_{col}, and R_{frame} are row number, column number and frame rate, respectively.

A straightforward approach to realize high speed image sensors is to arrange the output ports in parallel; however, it consumes more power and space owing to the large number of high speed output ports. Thus, this approach is applicable to consumer applications only if the number of parallel ports is small.

The other effective approach is the column-parallel processing [382, 383], which is suitable for CMOS image sensor architecture as described in Sec. 4.1. This approach is applicable to consumer electronics, such as smart phone cameras. The column-parallel processing can achieve a high data rate with a relatively slow processing time in a column, for example, about 1 μs. In this case, the speed is limited by the column parallel speed. Pixel-parallel processing is more effective for high-speed imaging than column parallel processing; however, it reduces the FF and hence, deteriorates the sensitivity. These are common features of pixel-parallel processing as described in Sec. 4.1. This drawback can be alleviated by introducing stacked technology or 3D technology described in Sec. 3.3.4 [273, 274, 384] of Chapter 3.

The last category is ultra-high-speed (UHS) CMOS image sensors. In this book, CMOS image sensors with over 10 kfps and 10 Gpixel/s are considered as UHS CMOS image sensors. The applications of UHS cameras are conventionally in scientific and industrial areas; however, recently the other important consumer application has emerged, namely the application for time-of-flight (TOF) cameras, which are described in Sec. 4.6.2.

A simple method for UHS CMOS image sensors is to limit the imaging area or to set the ROI (region-of-interest). This method can increase the frame rate by decreasing the number of pixels. To set the ROI, the random access approach discussed in Sec. 2.6.1 of Chapter 2 is employed. In this section, CMOS image sensors with full frame capturing are introduced instead of sensors with ROI.

To observe UHS phenomenon over 10 kfps and 10 G pixel/s, it is difficult to apply column- or pixel-parallel processing approach with the conventional pixel architecture, because R_{pix}= 10 G pixel/s requires a total readout time of about 100 ps. Thus, specialized architectures are required to observe such UHS phenomena.

In this section, global shutter pixels are introduced first; then CMOS image sensors with column-parallel and pixel-parallel processing are mentioned; and finally, UHS CMOS image sensors are described.

4.3.2 Global shutter

Because a CMOS image sensor is conventionally operated in the rolling shutter mode as described in Sec. 2.10, it is required to correctly acquire the image of a high-speed moving object; in other words, a global shutter (GS) is required [385–397]. Figure 4.8 shows a few types of GS pixels.

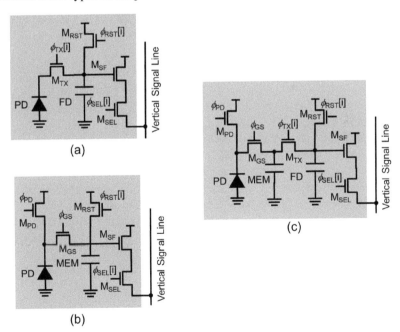

FIGURE 4.8: Basic pixel circuits for global shutter function: (a) Conventional 4T-APS without GS function; (b) 5T-global shutter APS; (c) 6T-global shutter APS. [i] denotes a local operation, while the others mean global operation.

The basic global shutter pixel consists of a 4T-APS plus PD shutter transistor M_{PD} as shown in Fig. 4.8 (b). The M_{PD} is also used for an OFD. In the figure, a conventional FD is used for a memory (MEM) node. The operating sequence is as follows. Firstly, the PD shutter transistor M_{PD} is turned on to reset the PD as the initial condition. As the PD is a pinned PD, all the carriers in the PD are transferred. After the integration time ends, the stored charges are transferred to the MEM by turning on the M_{GS}. This action is globally carried out. The stored charges in the MEM are sequentially read in row-by-row. In this type, the CDS operation cannot be introduced and therefore, the SNR is lower than that of 4T-APS. In addition, the stored charges in the MEM must be held for at least one frame time, but carriers generated by light unintentionally incident on the MEM region may affect the stored charges. Such a phenomenon in GS pixel is called parasitic light sensitivity (PSL). Furthermore, the dark current in the MEM deteriorates the signal. These are caused because FD is used as MEM.

To solve the above problems, the GS pixel as shown Fig. 4.8(c), has been developed. In the pixel, MEM is provided independently of FD, and one more transistor, that is, a storage gate transistor M_{SG} is required to transfer the charges stored in the PD to the MEM globally. The structure of the MEM is basically that of a pinned photodiode (PPD); however, some part of the diode is covered by gate. Thus, the potential under the gate area can be controlled by the applied voltage to the gate. Usually, to ensure the light shield, a metal, such as tungsten is used to cover the MEM.. In addition, the technology to reduce smear in CCD image sensors has been developed [398].

Figure 4.9 (a) shows the structure of 6T-type GS pixel and Figure 4.9 (b) shows its sequence of charge detection. The sequence is as follows: First, the photo-generated charges are accumulated in PD (step (1) in Fig. 4.9); then, the charges are transferred to the MEM region (step (2)); and stored in the MEM (step (3)). It should be noted that the MEM is divided into two regions. One region is where the storage gate covers the surface, and the other region is without the storage gate on the surface. In step (2), the photo-generated charges are transferred to the left region of the MEM by turning on the storage gate ϕ_{SG}. Next, in step (3), the transferred charges are stored in the MEM with ϕ_{SG} is turned off. As mentioned above, the MEM has the pinned PD structure; therefore, the dark current is extremely small and is also shielded by a metal of Tungsten to reduce the PSL. The final step (4) is the readout process where CDS is carried out. Before transferring the charges in the MEM, the FD is reset; the reset value is read out; then, the charges are transferred into the FD and the values are read out as conventional CDS process.

Steps (1) to (3) are global operation, while step (4) is local or row operation. The total timing chart including global and row operations is shown in Fig. 4.9 (c). In the GS pixel, the PD area is reduced so that the full well capacity decreases. This causes shrink in the dynamic range. To alleviate the issue, the pixel sharing technology described in Chapter 2, Sec. 2.5.3.2 is effective. In addition, the stacked technology described in Chapter 3, Sec. 3.3.4 is effective for GS CMOS image sensors [273].

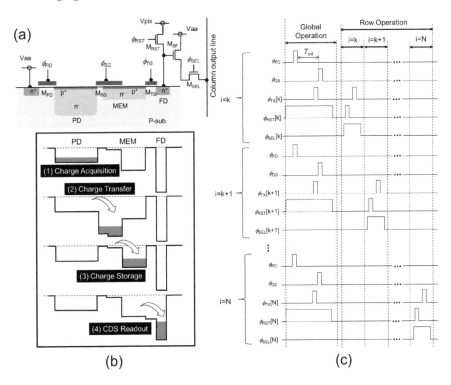

FIGURE 4.9: 6T-type GS pixel: (a) structure, (b) sequence of charge detection, and (c) timing chart. Illustrated after [387].

4.3.3 Column- and pixel-parallel processing for high speed imaging

By introducing column-parallel processing, CMOS image sensors can be made to achieve high frame rate more than 1 kfps or high pixel rate of more than 1 Gpixel/s [236, 383, 399–404]. Especially, image sensors for as high-definition television (HDTV) need high pixel rate [402]. For example, the CMOS image sensor with the 4K format (See Appendix G, about 8M pixels) and 120 fps needs a pixel rate of about 1 G pixels/s. However, if the column parallel processing is introduced, the column processing can be completed in about 4 μs, which is an acceptable processing time for one column. In addition, if the odd and even columns are separately output, then, the processing time is twice or about 8 μs.

Recently, the stacked CMOS image sensor technology has emerged and has been applied to high-speed CMOS image sensors. In these sensors, ADC has been employed in each pixel [273,274,384]. Therefore, the limitation is alleviated to more extent than with the column-parallel processing. For example, the pixel rate in [274] reached 7.96 Gpixels/s. The advancement of stacked technology may improve the pixel rate of the sensor.

4.3.4   Ultra-high-speed

For some industrial and scientific applications, ultra-high-speed (UHS) cameras with over 10 kfps are needed [405, 406]. The typical use for these cameras involves observing high-speed phenomena associated with combustion, shock waves, discharges, etc. The other use is a TOF camera, which is mentioned in Sec. 4.6.2 in this chapter. To meet this requirement, several methods of UHS image sensors have been developed. There are several factors that limit the UHS imaging as shown in Fig. 4.10 [405]. These are:

- T_{tr}: total transfer time of photo-generated carriers from the PD to the output signal line.
- T_{sel}: addressing time of a pixel, for example, settling time to turn on the select transistor.
- T_{sig}: addressing time of the signal line or the vertical line.
- T_{col}: column processing time.
- T_{read}: read-out time.

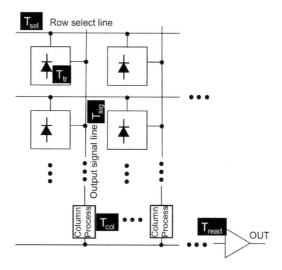

FIGURE 4.10: Limitation factors for ultra-high-speed CMOS image sensors: T_{tr} is the total transfer time of the photo-generated carriers from the PD to the output signal line; T_{sel} is the time required to address a pixel, for example, settling time to turn on the select transistor; T_{sig} is the time to address the signal line or the vertical line; T_{col} is the time for column processing; and T_{read} is the time to read out the output signal. Illustrated after [405].

Among the above limitation factors, the read-out time is critical in many cases. To alleviate this limitation, and to achieve UHS imaging, some of the UHS image sensors have burst mode besides the continuous mode as in conventional image sensors [407]. In the burst mode, signals from the pixels are stored in an analog memory array and the signals are output outside the sensor after the completion of the whole readout process. The transferring time to the memory is so short that the memory array is usually placed on a chip. The on-chip memory for UHS image sensors is classified into three types: the first one is the in-pixel memory [197, 407, 408]; the second one is the memory array associated with the pixel [409–411]; and the third one is the analog frame memory placed outside the image plane [412–414]. The pixel-associated memory is usually installed in the CCD image sensors, because a CCD can be inherently used as a memory array. It should be noted that the CCD was initially invented as a memory device [21]. The length of the memory in the UHS image sensors determines the number of frames. The frame memory outside image plane can be implemented with a long memory length compared to the other types of memory, although they consume the same area in the pixel. On the other hand, T_{sig} and T_{read} are critical parameters in the frame memory type. Figure 4.11 shows a UHS image sensor with an on-chip analog frame memory. In the sensor, a CDS and a current load for a source follower are integrated in a pixel to minimize T_{sig} and T_{read}.

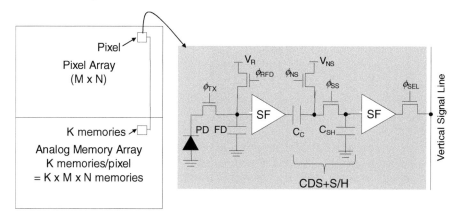

FIGURE 4.11: The UHS CMOS image sensor with on-chip analog frame memory array. Illustrated after [412, 414].

It is essential for such UHS image sensors to achieve photodetection with a high-speed response. In addition, high sensitivity is critical for high-speed CMOS image sensors, because the number of photons are incident in the conventional photodiodes, and there are no lateral electric fields. Therefore, the photo-generated carriers are not drifted but diffused, and thus the response speed is not very high. To accelerate the speed, it is essential to introduce an electric field inside photodetectors. A typical example is the avalanche photodiodes (APDs). SPAD is a type of APD that can be fabricated in a standard CMOS process. The detailed structure and characteristics are described in Sec. 3.2.2.3 of Chapter 3.

The other method is to introduce a lateral electric field in the photodiodes, especially the pinned photodiodes that are used in 4T-APS to achieve high SNR .

For UHS CMOS image sensors, a PD with a large area size is required to secure a sufficient number of photo-generated carriers even with a short exposure time. In such a large PD, the drift mechanism inside the PD area is more important for high speed response. The transit time for diffusion mode, t_{diff} is expressed as [415]

$$t_{\text{diff}} \approx \frac{l^2}{D_n},\tag{4.2}$$

where l and D_n are pixel size (square shape) and diffusion coefficient of electron, respectively. On the other hand, the transit time for the drift mode, t_{diff} is expressed as

$$t_{\text{drift}} \approx \frac{l^2}{\mu_n V_{\text{pinning}}},\tag{4.3}$$

where μ_n and V_{pinning} are the electron mobility and the pinning voltage of the PD, respectively. From Eqns. 4.2 and 4.3, and the Einstein relation below

$$D_n = \frac{k_B T}{e} \mu_n,\tag{4.4}$$

where k_B and T are the Boltzmann constant and the absolute temperature, respectively, the following equation is obtained.

$$\frac{t_{\text{diff}}}{t_{\text{drift}}} \approx \frac{\mu_n V_{\text{pinning}}}{D_n} \approx \frac{e V_{\text{pinning}}}{k_B T}.\tag{4.5}$$

As $k_B T$ is about 25 meV at room temperature and V_{pinning} is the range of 1-2 volts, the drift transit time is about two orders of magnitude faster than the diffusion transit time.

By introducing a lateral field, the photo-generated carriers are quickly transferred to the FD, even in a large-size PD. Two methods are shown below. One is to modify the shape and concentration of n-type region so as to form the distribution of fully depleted potential, which causes an electric field in the lateral direction. Figure 4.12 shows the PD structure where the lateral electric field is introduced by controlling the pinning voltage through changing the width of the PD [415].

The pinning voltage in the pinned PD is described in Sec. 2.4.3.1 of Chapter 2. In the PD, the vertical electrical field is also introduced with the gradient doping profile. The image sensor with the PD structure is originally designed for the TOF sensor. The UHS CMOS image sensor in Fig. 4.11 has introduced a similar mechanism for lateral electric field in the PD as shown in Fig. 4.13 [412, 414]. Because in this sensor, the area size of the photodiode is large with dimensions of 30.00 μm $\times$ 21.34 μm, changing the width of the n-layer like in [415] is introduced with several directions, as well as introducing three different doping concentrations radially.

The other UHS image sensor is to introduce four control electrodes to enhance in-plane electric field as shown in Fig. 4.14 [188]. Since the control electrodes are arranged on both sides of the PD, the electrodes do not cover the PD surface, and the FD can be arranged normally, compared with similar structure shown in Fig. 3.5. This type of UHS image sensors is more suitable to high speed modulation and thus suitable to TOF sensors described in Sec. 4.6.2.

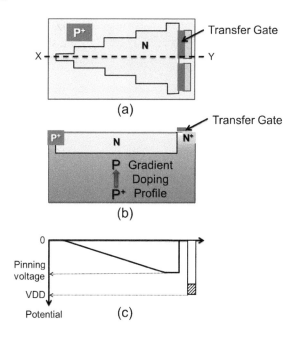

FIGURE 4.12: The PD structure in the UHS CMOS image sensor. (a) Top view, (b) cross-sectional view, and (c) potential profile. Illustrated after [415]. By courtesy of Dr. C. Tubert.

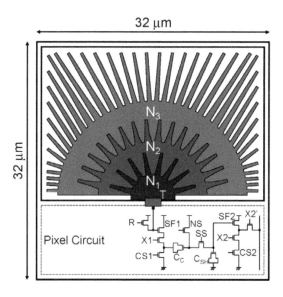

FIGURE 4.13: The PD structure in the UHS CMOS image sensor and its pixel circuits. Adapted from [414] with permission.

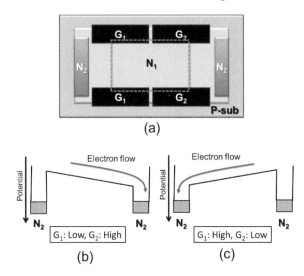

FIGURE 4.14: The UHS CMOS image sensor with lateral electric field charge modulator. Illustrated after [188].

4.4 Wide dynamic range

4.4.1 Overview

The human eye has a wide dynamic range of about 200 dB. To achieve such a wide dynamic range, the eye has three mechanisms [416] as shown in Appendix C. First, the human eye has two types of photo-receptor cells, cones and rods which correspond to two types of photodiodes with different photo-sensitivities. Second, the response curve of the eye's photo-receptor is logarithmic so that saturation occurs slowly. Third, the response curve shifts according to the ambient light level or averaged light level. Conventional image sensors, in contrast, have a dynamic range of 60–70 dB, which is mainly determined by the well capacity of the photodiode.

Some applications, such as those in automobiles and for security, require a dynamic range over 100 dB [369]. To expand the dynamic range, many methods have been proposed and demonstrated. They can be classified into four categories as shown in Table 4.1: dual sensitivity, nonlinear response, multiple sampling, and saturation detection. Figure 4.15 illustrates nonlinear, multiple sampling and saturation detection methods. An example image taken by a wide dynamic range (WDR) image sensor developed by S. Kawahito and his colleagues at Shizuoka Univ. [417] is shown in Fig. 4.16. The extremely bright light-bulb at the left can be seen, as well as the objects under a dark condition to the right.

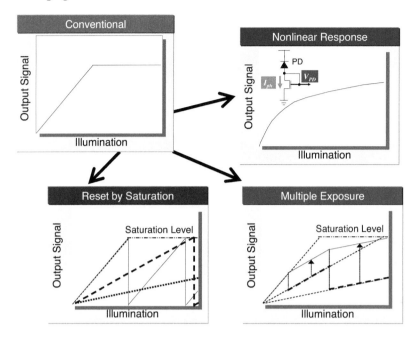

FIGURE 4.15: Basic concept of enhancing dynamic range.

TABLE 4.1

Smart CMOS sensors for a wide dynamic range

Category	Implementation method	Refs.
Dual sensitivity	Dual PDs in a pixel	[418, 419]
	Dual accumulation capacitors in a pixel	[420–422]
Nonlinear response	Log sensor	[174–177, 179, 423]
	Log/linear response	[424–426]
	Well capacity adjustment	[427–431]
	Control integration time/gain	[432, 433]
Multiple sampling	Dual sampling	[434–436]
	Multiple sampling with fixed short exposure time	[437, 438]
	Multiple sampling with varying short exposure time	[417]
	Multiple sampling with pixel-level ADC	[155, 198, 439]
Saturation detection	Locally integration time and gain	[377]
	Saturation count	[377, 440–444]
	Pulse width modulation	[445]
	Pulse frequency modulation	[156, 206, 207, 213, 446, 447]

FIGURE 4.16: Example image taken by a wide dynamic range image sensor developed by S. Kawahito *et al.* [417]. The image is produced by synthesizing several images; the details are given in Sec. 4.4.3.2. By courtesy of Prof. Kawahito at Shizuoka Univ.

As mentioned above, the human retina has two types of photo-receptors with high and low sensitivities. An image sensor with two types of PDs with high and low sensitivities can achieve a wide dynamic range. Such an image sensor has already been produced using CCDs [448]. CMOS image sensors with dual PDs have been reported in Refs. [418] and [419] that also have two types of photo-detectors with high and low sensitivities.

Nonlinear response is a method to modify the photo-response from linear to nonlinear, for example, a logarithmic response. This method can be divided into two methods, using a log sensor and well-capacity adjustment. In a log sensor, a photodiode has a logarithmic response. By adjusting the well capacity, the response can be changed to be nonlinear, but in some cases, a linear response is achieved, which is mentioned later. By using overflow drain (OFD) as the second accumulation capacitance, nonlinear response can be achieved. Nonlinear response may cause more complicated color processing compared with linear response.

Multiple sampling is a method where the signal charges are read several times. This is a linear response. For example, bright and dim images are obtained with different exposure times and then the two images are synthesized so that both scenes can be displayed in one image. Extending the dynamic range by well capacity adjustment and multiple sampling is analyzed in detail in Ref. [449].

In the saturation detection method, the integration signal or accumulation charge signal is observed and if the signal reaches a threshold value, then, the accumulation charge is reset and the reset number is counted. Repeating this process, the final output signal is obtained for the residue charge signal and the reset count number. There are several variations to the saturation detection method. Pulse modulation is one alternative and is discussed in Sec. 3.2.2 of Chapter 3. In this method, the integration time is different from pixel to pixel. For example, in PWM, the output is the pulse width or counting values so that the maximum detectable light intensity is determined by the minimum countable value or clock, and the minimum detectable value is determined by the dark current. Thus the method is not limited by the well capacity, so that it has a wide dynamic range. In the following sections, examples of the above methods are described.

4.4.2 Nonlinear response

4.4.2.1 Log sensor

The log sensor is described in Chapter 3, Sec. 3.2.1.2. It is used for wide dynamic range imagers, and can now obtain a range of over 100 dB [174–177, 179, 423].

Issues of the log sensor include variations of the fabrication process parameters, a relatively large noise in low light intensity, and image lag. These disadvantages are mainly exhibited in sub-threshold operation, where the diffusion current is dominant.

There have been some reports of the achievement of logarithmic and linear responses in one sensor [424–426]. The linear response is preferable in the dark light region, while the logarithmic response is suitable in the bright light region. In these sensors, calibration in the transition region is essential.

4.4.2.2 Well-capacity adjustment

The well-capacity adjustment is a method to control the well depth or full well capacity in the charge accumulation region during integration. The full well capacity is described in Sec. 2.5.2.1 of Chapter 2. In this method, a drain is used for the overflow charges. Controlling the gate between the accumulation region and the overflow drain (OFD) over time, $\phi_b(t)$, the photo-response curve becomes nonlinear.

Figure 4.17 illustrates the 4T-APS based pixel structures with an OFD to enhance the maximum well capacity. The sensitivity in a 4T-APS is better than in a 3T-APS and therefore the dynamic range from the dark light condition to the bright light condition can be improved. When strong illumination is incident on the sensor, the photo-generated carriers are saturated in the PD well and flow over into the OFD capacitance. In 4T-APS, the photo-generated carriers over the PD well capacity are transferred into the FD. In Fig. 4.17 (a), the gate voltage of the reset transistor M_{RS} is varied over time as $\phi_b(t)$ [429], while in Fig. 4.17 (b), the gate voltage of the transfer transistor M_{TX} is varied over time as $\phi_b(t)$ [430]. By decreasing the potential well gradually as $\phi_b(t)$, strong light intensity can hardly saturate the well and also weak light can be detected. Usually, $\phi_b(t)$ is changed like stepwise, the photo-response curve shows multiple knees. The drawback of this method is that the overflow mechanism consumes the pixel area and hence, the FF is reduced.

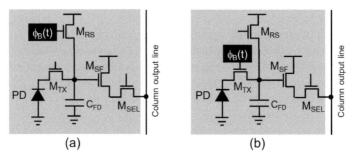

FIGURE 4.17: 4T-APS based pixel structures for wide dynamic range with well-capacity adjustment. The well-capacity adjustment signal $\phi_b(t)$ is applied to the gate of the reset transistor M_{RS} (a) [429], and to the gate of the transfer transistor M_{TX} (b). Illustrated after [430].

4.4.3   Linear response

4.4.3.1   Dual sensitivity

Dual PD   When two types of photo-detectors with different sensitivities are integrated in a pixel, a wide range of illuminations can be covered; under bright conditions, the PD with lower sensitivity is used, while under dark conditions, the PD with higher sensitivity is used. This is very similar to the human visual system, as mentioned above, and has already been implemented using CCDs [448]. In the work reported in Ref. [450], an FD is used as a low sensitivity photo-detector. Under bright light conditions, some carriers are generated in the substrate and diffuse to the FD region, contributing to the signal. This structure was first reported in Ref. [451]. This is a direct method of detection, so that no latency of captured images occurs, in contrast with the multiple sampling method mentioned in Sec. 4.4.3.2. Another implementation of dual photo-detectors has been reported in Ref. [452], where a PG is used as the primary PD and an n-type diffusion layer is used as the second PD, and in Ref. [418], two pinned PDs are implemented in 4T-APS. The dual PD method has the disadvantage that the sensor sensitivity depends on the light incident angle because PDs with different sensitivities are in different locations in the pixel. In Ref. [419], by arranging two PDs concentrically, this drawback can be alleviated.

Dual accumulation   In the well-capacity adjustment, the adjustment signal $\phi_b(t)$ is varied over time to properly transfer the overflow charges. Consequently, the photo-response curve becomes non-linear. Sugawa *et al.* of Tohoku University have developed wide dynamic range 4T-APS type CMOS image sensors where the overflow carriers from FD are accumulated in another capacitance, C_S, that is a dual accumulation structure, as shown in Fig. 4.18 [420–422]. In this study, a stacked capacitor used for the lateral OFD (overflow drain) capacitance C_S was developed as shown in Fig. 4.18. It is called lateral overflow integration capacitor (LOIC). The photo-generated charges are transferred to the FD and C_S and accumulated in the two capacitors for saturated intensity and in only FD for non-saturated intensity. Thus, if the light intensity is weak, the photo-generated charges are stored only in FD, which is the same in conventional 4T-APS, while strong light intensity, the photo-generated charges are stored in both FD and C_S so that the accumulation capacitance increases. This effectively expands the well-capacity.

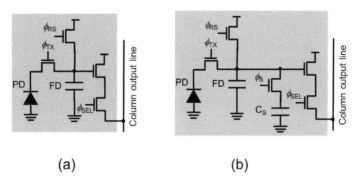

(a)   (b)

FIGURE 4.18: Lateral overflow integration capacitor. (a) Conventional 4T-APS. (b) 4T-APS with Lateral overflow integration capacitor. Illustrated after [420].

By introducing a direct photocurrent output mode in the above structure, an ultra-high dynamic range of over 200 dB has been achieved, as reported in Ref. [421], by combining with the architecture reported in Ref. [420]. In the region of the direct photo-current output mode, a logarithmic response is employed.

In Ref. [431], a 3T-APS and a PPS are combined in a pixel to enhance the dynamic range. Generally, an APS has a superior SNR at low illumination compared with a PPS, suggesting that at bright illumination, a PPS is acceptable. A PPS is suitable for use with an OFD because a column charge amplifier can completely transfer charges in both the PD and OFD. In this case, care need not be taken with regard to whether or not signal charges in PD are transferred into the OFD.

4.4.3.2 Multiple sampling

Multiple sampling is a method to read signal charges several times and synthesize those images in one image. This method is simple and easily achieves a wide dynamic range. However, it has an issue regarding synthesis of the images obtained.

Dual sampling   In dual sampling, two sets of readout circuits are employed in a chip [434, 436]. When pixel data in the n-th row are sampled and held in one readout circuitry and then reset, pixel data in the $(n - \Delta)$-th row are sampled and held in another readout circuitry and reset. The array size is $N \times M$, where N is the number of rows and M is the number of columns. In this case, the integration time of the first readout row is $T_l = (N - \Delta)T_{row}$ and the integration time of the second readout row is $T_s = \Delta T_{row}$. Here, T_{row} is the time required to readout one row of data and $T_{row} = T_{SH} + MT_{scan}$, where T_{SH} is a sample and hold (S/H) time and T_{scan} is the time to read out data from the S/H capacitors or the scanning time. It is noted that by using the frame time $T_f = NT_{row}$, T_l is expressed as

$$T_l = T_f - T_s. \tag{4.6}$$

The dynamic range is given by

$$DR = 20\log\frac{Q_{max}T_l}{Q_{min}T_s} = DR_{org} + 20\log\left(\frac{T_f}{T_s} - 1\right). \tag{4.7}$$

Here, DR_{org} is the dynamic range without dual sampling and Q_{max} and Q_{min} are the maximum and minimum accumulation charges, respectively. For example, if $N = 480$ and $\Delta = 2$, then the ratio of the accumulation $T_f/T_s \approx T_l/T_s$ becomes about 240, so that it can expand the dynamic range of about 47 dB.

This method only requires two S/H regions and no changes in pixel structure; therefore, it can be applied, for example, to a 4T-APS, which has a high sensitivity. The disadvantages of this method are that only two accumulation times are obtained and a relatively large SNR dip is exhibited at the boundary of the two different exposures.

At the boundary of the two different exposures, the accumulated signal charges change from maximum value Q_{max} to $Q_{max}T_s/T_l$, which causes a large SNR dip. The SNR dip ΔSNR is

$$\Delta SNR = 10\log\frac{T_s}{T_l}. \tag{4.8}$$

If the noise level is not changed in the two regions, ΔSNR is equal to ≈ -24 dB for the above example ($T_l/T_s \approx 240$).

Fixed short time exposure   To reduce the SNR dip, M. Sasaki *et al.* introduced multiple short time exposures [437, 438]. In non-destructive readout, multiple sampling is possible. By reading the short time integration T_s a total of k times, the SNR dip becomes

$$\Delta SNR = 10\log k\frac{T_s}{T_l}. \tag{4.9}$$

It is noted that in this case T_l is expressed as

$$T_l = T_f - kT_s. \tag{4.10}$$

Thus the dynamic range expansion ΔDR is

$$\Delta DR = 20\log\left(\frac{T_f}{T_s} - k\right), \tag{4.11}$$

which is slightly different from T_f/T_s. If $T_f/T_s = 240$ and $k = 8$, then $\Delta DR \approx 47$ dB and $\Delta SNR \approx -15$ dB.

Varying short exposure times   In the previous method, a short exposure time was fixed and read several times. M. Mase *et al.* have improved the method by varying the short exposure time [417]. In the short exposure time slot, several exposure times were employed. During the readout time of one short exposure period, a shorter exposure period is inserted, while a further shorter exposure period is inserted and so on. This is illustrated in Fig. 4.19. With a fast readout circuitry with column parallel cyclic ADCs, it is possible to realize this method. In this method, the dynamic range expansion ΔDR is

$$\Delta DR = 20\log \frac{T_l}{T_{s,min}}, \tag{4.12}$$

where $T_{s,min}$ means the minimum exposure time. In this method, the SNR dip can be reduced by minimizing each exposure time for the ratio of T_l to Ts.

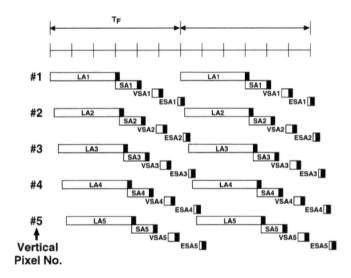

FIGURE 4.19: Exposure and readout timing for multiple sampling with varying short exposure times. LA: long accumulation, SA: short accumulation, VSA: very short accumulation, ESA: extremely short accumulation. Adapted from [417] with permission.

Pixel-level ADC   Another multiple sampling method is to implement pixel-level ADCs [155,198,439]. In this method, a bit-serial ADC with single slope is employed for every four pixels. According to the integration time, the precision of the ADC is varied to obtain a high resolution; for a short integration time, a higher precision ADC is executed. The image sensor reported in Ref. [198] is integrated with a DRAM frame memory in the chip and achieves a dynamic range over 100 dB with a pixel number of 742×554.

4.4.3.3   Saturation detection

The saturation detection method is based on monitoring and controlling the saturation signal. The method is asynchronous and hence, automatic exposure of each pixel is easily achieved. A common issue with this method is suppression of the reset noise; it is difficult to employ a noise cancellation mechanism due to the multiple reset action. A decreased SNR in the reset is also an issue.

Saturation count When the signal level reaches the saturation level, the accumulation region is reset and accumulation restarts. By repeating the process and counting the number of resets in a time duration, the total signal charge in the time period can be calculated with the residue charge signal and the number of resets [377, 440–444]. The counting circuitry is implemented at the pixel level [440, 441], and in a column level [377, 442, 443]. At the pixel level, the FF is reduced due to the extra area required for the counting circuitry. For either the pixel level or column level, frame memory is required in this method. This method is essentially the same as PFM.

Pulse width modulation is mentioned in Sec. 3.2.2.1 and is used for wide dynamic range image sensors [192].

Pulse frequency modulation Pulse frequency modulation (PFM) is discussed in Sec. 3.2.2.2 and is used for wide dynamic range image sensors [156, 204–206].

4.5   Demodulation

4.5.1   Overview

In the demodulation method, a modulated light signal is illuminated on an object and the reflected light is acquired by the image sensor. The method is effective in detecting signals with a high SNR, because it enables the sensor to detect only the modulated signals and thus removes any static background noise. Implementing this technique in an image sensor with a modulated light source is useful for such fields as ADAS, factory automation (FA), and robotics, because such a sensor could acquire an image, while being minimally affected by background light conditions. In addition to such applications, this sensor could be applied to tracking a target specified by a modulated light source. For example, motion capture could be easily realized by this sensor under various illumination conditions. Another important application of the demodulation technique is the 3D range finder with time-of-flight (TOF) method, as described in Sec. 4.6.

It is difficult to realize demodulation functions in conventional image sensors, because a conventional image sensor operates in the accumulation mode and hence, the modulated signal is washed out by the accumulating modulation charges. The lock-in pixel described in Sec. 3.2.1.4 can be applied to the demodulation functions. The concept of demodulation technique in a smart CMOS image sensor is illustrated in Fig. 4.20. The illumination light $I_o(t)$ is modulated by frequency f, and the reflected (or scattered) light $I_r(t)$ is also modulated by f. The sensor is illuminated by the reflected light $I_r(t)$ and the background light I_b, that is, the output from the photodetector is proportional to the sum $I_r(t) + I_b$. The output from the photodetector is multiplied by the synchronous modulated signal $m(t)$ and then integrated. Thus the output V_{out} is produced [453]

$$V_{out} = \int_{t-T}^{t} (I_r(\tau) + I_b) m(\tau) d\tau, \tag{4.13}$$

where T is the integration time.

There are several reports on realizing a demodulation function in smart CMOS image sensors based on the concept shown in Fig. 4.20. Its implementation method can be classified into two categories: the correlation method [175, 454–457], and the method using two accumulation regions in a pixel [458–463]. The lock-in pixel is a method of this type, which is described in Sec. 3.2.1.4 of Chapter 3. The correlation method is a straightforward implementation of the concept in Fig. 4.20.

4.5.2 Correlation

The correlation method is based on multiplying the detected signal with the reference signal and then integrating it or performing a low-pass-filtering. The process is described by Eq. 4.13. Figure 4.21 shows the concept of the correlation method. The key component in the correlation method is a multiplier. In Ref. [457], a simple source connected type multiplier [44] is employed, while in Ref. [175] a Gilbert cell [134] is employed to subtract the background light.

In this method, three-phase reference is preferable to obtain sufficient modulation information on the amplitude and phase. Figure 4.22 shows the pixel circuits to implement the three-phase reference. The source connected circuits with three

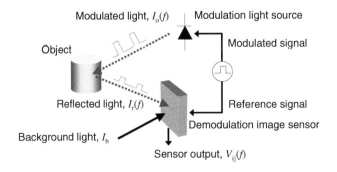

FIGURE 4.20: Concept of a demodulation image sensor.

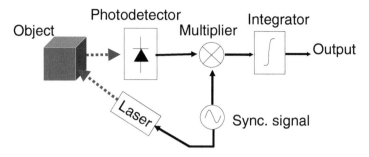

FIGURE 4.21: Conceptual illustration of the correlation method.

reference inputs are suitable for this purpose [457]. This gives amplitude modulation (AM)–phase modulation (PM) demodulation.

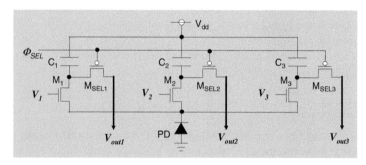

FIGURE 4.22: Pixel circuit for the correlation method. Illustrated after [457].

In the circuit, the drain current I_i in each M_i is expressed as

$$I_i = I_o exp\left(\frac{e}{mk_BT}V_i\right).$$

(4.14)

The notations used in the equation are the same as those in Appendix F. Each drain current of M_i, I_i is then given by

$$I_i = I\frac{\exp\left(-\frac{eV_i}{mk_BT}\right)}{\exp\left(-\frac{eV_1}{mk_BT}\right)+\exp\left(-\frac{eV_2}{mk_BT}\right)+\exp\left(-\frac{eV_3}{mk_BT}\right)}.$$

(4.15)

The correlation equations are thus obtained as

$$I_i - I/3 = -\frac{e}{3mk_BT}I(V_i - \bar{V}),$$

(4.16)

where $\bar{V}$ is the average of V_i, $i = 1, 2, 3$.

This method has been applied to a 3D range finder [460, 464] and to spectral matching [465].

4.5.3   Method of two accumulation regions

The method of using two accumulation regions in a pixel is essentially based on the same concept, but with a simpler implementation, as shown in Fig. 4.23. The method is basically similar as one in the lock-in pixel described in Sec. 3.2.1.4 of Chapter 3. In this method, the modulated signal is a pulse or ON–OFF signal. When the

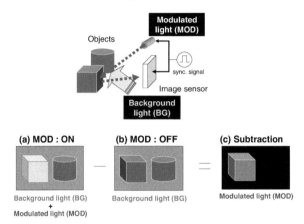

FIGURE 4.23: Concept of the two accumulation regions method.

modulated signal is ON, the signal accumulated in one of the accumulation regions. Correlation is achieved by this operation. When the modulated signal is OFF, the signal accumulates in the other region and the background signal can be removed by subtracting the two signals.

Figure 4.24 shows the pixel structure with two accumulation regions [458]. Figure 4.25 shows the pixel layout using a 0.6-μm 2-poly 3-metal standard CMOS technology. The circuit consists of a pair of readout circuits similar to a conventional photogate (PG) type APS. In other words, two transfer gates (TX1 and TX2) and two floating diffusions (FD1 and FD2) are implemented. One PG is used as a photodetector instead of a photodiode, and is connected with both FD1 and FD2 through TX1 and TX2, respectively. The reset transistor (RST) is common to the two readout circuits. The two outputs OUT1 and OUT2 are subtracted from each other, and thus only a modulated signal is obtained. A similar structure with two accumulation regions is reported in Ref. [463].

The timing diagram of this sensor is shown in Fig. 4.26. First, the reset operation is achieved by turning the RST on when the modulated light is OFF. When the modulated light is turned on, the PG is biased to accumulate photocarriers. Then, the modulated light and the PG are turned off to transfer the accumulated charges to FD1 by opening TX1. This is the ON-state of the modulated light; in this state, both the modulated and the static light components are stored in FD1. Next, the PG is biased again and starts to accumulate charges in the OFF-state of the modulated light. At the end of the OFF period of the modulated light, the accumulated charges

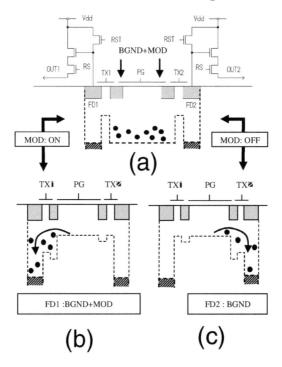

FIGURE 4.24: Pixel structure of a demodulated CMOS image sensor. The accumulated charges in the PG (a) are transferred to FD1 when the modulation light is ON (b) and transferred to FD2 when the modulation light is OFF (c).

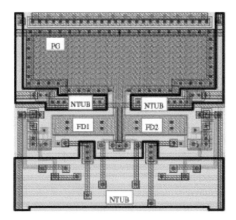

FIGURE 4.25: Pixel layout of the demodulated image sensor. The pixel size is $42 \times 42 \mu m^2$. Adapted from [459] with permission.

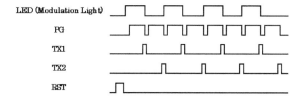

FIGURE 4.26: Timing chart of the demodulation image sensor [458]. PG: photogate, TX: transfer gate, RST: reset.

are transferred to FD2. Thus only the static light component is stored in FD2. By repeating this process, the charges in the ON- and OFF-states accumulate in FD1 and FD2, respectively. According to the amount of accumulated charge, the voltages in FD1 and FD2 decrease in a stepwise manner. By measuring the voltage drops of FD1 and FD2 at a certain time and subtracting them from each other, the modulated signal component can be extracted.

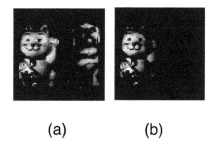

(a)   (b)

FIGURE 4.27: (a) Normal and (b) demodulated images. Adapted from [459] with permission.

Figure 4.27 shows experimental results obtained using the sensor [456]. One of the two objects (a cat and a dog) is illuminated by modulated light and the demodulated image only shows the cat, which is illuminated by the modulated light. Figure 4.28 shows further experimental results. In this case, a modulated LED is attached in the neck of the object (a dog), which moves around. The demodulated images show only the modulated LED and thus give a trace of the object. This means that a demodulation sensor can be applied to target tracing. Another application for a camera system suppressing saturation is presented in Ref. [466].

This method achieves an image with little influence from the background light condition. However, the dynamic range is still limited by the capacity of the accumulation regions or full well capacity. In Ref. [460, 461], the background signal is subtracted in every modulation cycle so that the dynamic range is expanded.

Although the adding circuits consume pixel area, this technique is effective for demodulation CMOS image sensors.

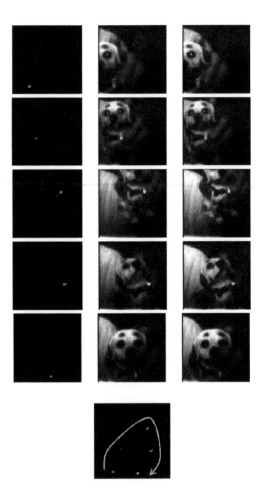

FIGURE 4.28: Demonstration of marker tracking. The images are placed in order of time from top to bottom. Left column: modulated light pattern extracted by the sensor. Middle column: output from the modulated light and background light. Right column: output from only the background light. The bottom figure shows the moving trace of the marker. For convenience, the moving direction and the track of the LED are superimposed. Adapted from [459] with permission.

4.6 Three-dimensional range finder

4.6.1 Overview

A range finder is an important application for factory automation (FA), ADAS, robot vision, gesture recognition, etc. A LIDAR (light detection and ranging) is one type of range finders. By using smart CMOS image sensors, three-dimensional (3D) range finding or image acquisition associated with distances can be realized. Several approaches suitable for CMOS image sensors have been investigated. Their principles are based on time-of-flight (TOF), triangulation, and other methods summarized in Table 4.2. Figure 4.29 shows the concept of three typical methods: TOF, binocular and light section. Figure 4.30 shows images taken with a 3D range finder [467]. The distance to the object is shown on the image.

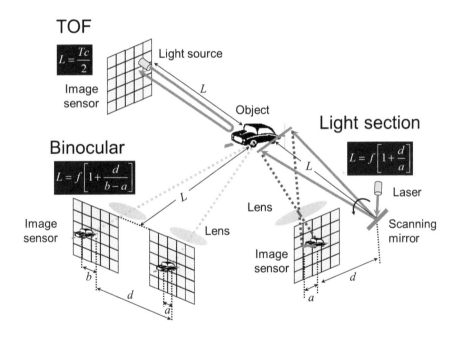

FIGURE 4.29: Concept of three methods for 3D range finders: TOF, binocular, and light section.

TABLE 4.2
Smart CMOS sensors for 3D range finders

Category	Implementation method	Affiliation and Refs.
Direct TOF	APD array	ETH [92, 93], MIT [468]
Indirect TOF	Pulse	ITC-irst [469], Fraunhofer [470], Sizuoka U. [471]
	Sinusoidal	PMD [472], CSEM [463], Canesta [473]
Triangulation	Binocular	Shizuoka U. [474], Tokyo Sci. U. [475], Johns Hopkins U. [476]
	Structured light	Carnegie Mellon U. [89], U. Tokyo [477], SONY [478], Osaka EC. U. [455]
Others	Light intensity (Depth of intensity)	Toshiba [479]

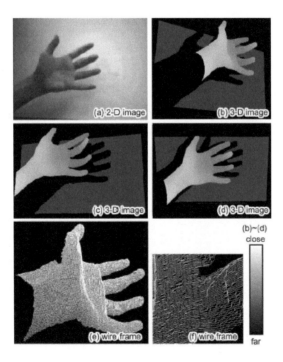

FIGURE 4.30: Examples of images taken by a 3D range finding image sensor [467]. Adapted from [467] with permission.

4.6.2 Time-of-flight

Time-of-flight (TOF) is a method to measure the round-trip time of flight and has been used in light detection and ranging (LIDAR) for many years [480]. The distance to the object L is expressed as

$$L = \frac{Tc}{2}, \tag{4.17}$$

where T is the round-trip time (TOF $= T/2$) and c is the speed of light. The most notable feature of TOF is its simplicity; it requires only a TOF sensor and a light source. TOF sensors are classified into direct and indirect TOF sensors.

4.6.2.1 Direct TOF

A direct TOF sensor measures the round-trip time of light in each pixel directly. Consequently, it requires a high speed photodetector and high precision timing circuits. For example, for $L = 3$ m, $T = 10^{-8}$s$= 10$ ps. To obtain accuracy in mm, an averaging operation is necessary. The advantage of direct TOF is its wide range for measuring distance, from meters to kilometers.

As high-speed photo-detectors in the standard CMOS technology, the APDs are used in the Geiger mode, i.e. the SPADs for direct TOF sensors [92, 93, 468], as discussed in Sec. 2.3.4. A TOF sensor with 32×32 pixels, each of which is integrated with the circuits shown in Fig. 2.11 in Sec. 2.3.4.1 with an area of 58×58 μm^2, has been fabricated in 0.8-μm standard CMOS technology with a high voltage option. The anode of the APD is biased at a high voltage of -25.5 V. The jitter of the pixel is 115 ps, and hence, to obtain mm accuracy an averaging operation is necessary. A standard deviation of 1.8 mm is obtained with multiple depth measurements with 10^4 at a distance of around 3 m.

4.6.2.2 Indirect TOF

To alleviate the limitations for the direct TOF, the indirect TOF has been developed [463, 469, 470, 472, 473, 481, 482]. In the indirect TOF, the round-trip time is not measured directly; instead, two modulated light signals are used. An indirect TOF sensor generally has two accumulation regions in each pixel to demodulate the signal, as mentioned in Sec. 4.5 or the lock-in pixel, as mentioned in Sec. 3.2.1.4 of Chapter 3. The timing diagram of indirect TOF is illustrated in Fig. 4.31. In this figure, two examples are shown.

When the modulation signal is a pulse or an on/off signal, two pulses with a delay time t_d between them are emitted with a repetition rate of the order of MHz. Figure 4.32 illustrates the operating principle [469, 481]. In this method, the TOF signal is obtained as follows. Two accumulation signals V_1 and V_2 correspond to the two pulses with the same width t_p, as shown in Fig. 4.31(a). From t_d and the two times

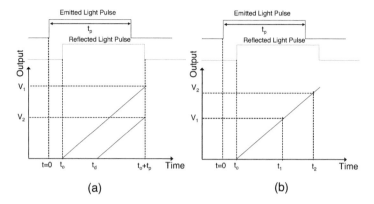

(a)   (b)

FIGURE 4.31: Timing diagram of indirect TOF with two different pulses. (a) The second pulse has a delay of t_d against the first pulse. (b) Two pulses with different durations. Illustrated after [469].

of the TOF, the distance L is computed as

$$t_d - 2 \times TOF = t_p \frac{V_1 - V_2}{V_1} \tag{4.18}$$

$$TOF = \frac{L}{c} \tag{4.19}$$

$$\therefore L = \frac{c}{2}\left[t_p\left(\frac{V_2}{V_1} - 1\right) + t_d\right]. \tag{4.20}$$

In Ref. [469], an ambient subtraction period is inserted in the timing and thus a good ambient light rejection of 40 Klux is achieved. A 50×30-pixel sensor with a pixel size of $81.9 \times 81.7 \ \mu m^2$ has been fabricated in a standard $0.35\text{-}\mu m$ CMOS technology and a precision of 4% is obtained in the 2–8 m range.

Another indirect TOF with pulse modulation uses two pulses with different timings, T_1 and T_2 [470, 483, 484]. The procedure to estimate L is as follows. V_1 and V_2 are output signal voltages for the shutter time t_1 and t_2, respectively. As shown in Fig. 4.31(b), from these four parameters, the intersect point $t_o = TOF$ can be interpolated. Consequently:

$$L = \frac{1}{2}c\left(\frac{V_2 t_1 - V_1 t_2}{V_2 - V_1}\right). \tag{4.21}$$

Next, a sinusoidal emission light is introduced instead of the pulse for the indirect TOF [463, 472, 473, 485].

The TOF is obtained by sampling four points, each of which is shifted by $\pi/2$, as shown in Fig. 4.33, and calculating the phase shift value ϕ [485]. The four sampled values A_1–A_4 are expressed by the signal phase shift ϕ, the amplitude a, and the

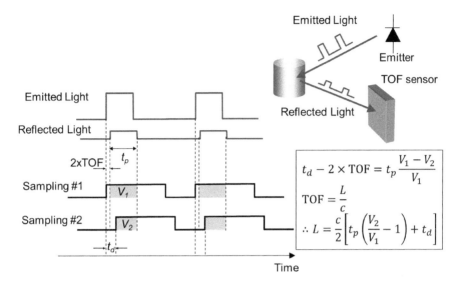

FIGURE 4.32: Operating principle of indirect TOF with two emission pulses with a delay.

offset b as

$$A_1 = a\sin\phi + b, \tag{4.22}$$
$$A_2 = a\sin(\phi + \pi/2) + b, \tag{4.23}$$
$$A_3 = a\sin(\phi + \pi) + b, \tag{4.24}$$
$$A_4 = a\sin(\phi + 3\pi/2) + b. \tag{4.25}$$

From the above equations, ϕ, a, and b are solved as

$$\phi = \arctan\left(\frac{A_1 - A_3}{A_2 - A_4}\right). \tag{4.26}$$

$$a = \frac{\sqrt{(A_1 - A_3)^2 + (A_2 - A_4)^2}}{2}. \tag{4.27}$$

$$b = \frac{A_1 + A_2 + A_3 + A_4}{4}. \tag{4.28}$$

Finally, by using ϕ, the distance L can be calculated as

$$L = \frac{c\phi}{4\pi f_{mod}}, \tag{4.29}$$

where f_{mod} is the repetition frequency of the modulation light.

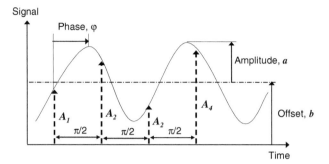

FIGURE 4.33: Operating principle of indirect TOF with sinusoidal emission light. Illustrated after [454].

To realize indirect TOF in a CMOS sensor, several types of pixels have been developed with two accumulation regions in a pixel. They are classified into two techniques: one is to place two FDs on either side of the photo-detector [463, 471, 473, 485], and the other is to use a voltage amplifier to store signals in capacitors [470, 481]. In other words, the first technique is to change the photo-detection device, while the second is to use conventional CMOS circuitry. The photomixing device (PMD) is a commercialized device that has a PG with two accumulation regions on either side [485].

4.6.3 Triangulation

Triangulation is a method to measure the distance to the field-of-view (FOV) by a triangular geometrical arrangement. This method can be divided into two classes: passive and active. The passive method is also called the binocular or stereo-vision method. In this method, two sensors are used. The active method is called the structured light method or light-section method. In this method a patterned light source is used to illuminate the FOV.

4.6.3.1 Binocular method

The passive method has the advantage that it requires no light source and only two sensors are needed. Reports of several such sensors have been published [474, 476]. The two sensors are integrated into two sets of imaging areas to execute stereo-vision. However, this means that the method must include a technique to identify the same FOV in the two sensors, which is a complicated problem for typical scenes. For the sensor reported in Ref. [474], the FOV is restricted to a known object and 3D information is used to improve the recognition rate of the object. The sensor has two imaging areas and a set of readout circuitry including the ADCs. The sensor in Ref. [476] integrates the current mode disparity computation circuitry.

By using diffractive optics, the right and left beams can be separated by a lenticular lens as a beam splitter and a digital micro-lens acts as focus enhancer. The digital micro-lens is based on a diffractive optics. The schematic structure is shown in Fig. 4.34 [326]. By using this configuration, parallax can be achieved, and thus 3-D imaging can be realized.

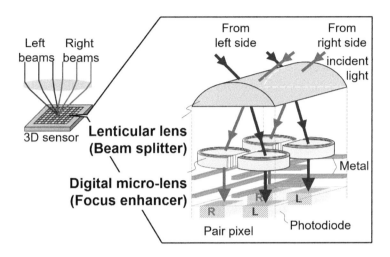

FIGURE 4.34: Schematic structure of multi-ocular image sensor with a beam splitting lens and digital micro-lens. By courtesy of Dr. S. Koyama.

4.6.3.2 Structured light method

In the structured light method or light section method, a structured light source is required, which is a two-dimensional pattern to illuminate target objects. For example, a stripe pattern is illuminated on an object so that the projected stripe pattern is distorted along the 3D shape of the object. To acquire the distorted image, the distance can be estimated based on the triangular method. Many projection patterns have been reported [486].

The multiple stripe pattern is obtained by scanning one stripe on an object. In this case, the light source is scanned over the FOV. Considerable research on such structured light methods have been published [168, 455, 460, 467, 477, 487]. To distinguish the projected light pattern from the ambient light, a high power light source with a scanning system is required. In Ref. [488], the first integrated sensor with 5×5 pixels was reported for 2-μm CMOS technology.

In the structured light method, finding a pixel at the maximum value is essential. In Refs. [467, 487], a conventional 3T-APS is used in two modes, the normal image output mode and the PWM mode, as shown in Fig. 4.35. PWM is discussed in Sec. 3.2.2. PWM can be used as a 1-bit ADC in this case. In the sensor, PWM-ADCs are located in the column and thus column-parallel ADC is achieved. In Fig. 4.35(b), the pixel acts as a conventional 3T-APS, while in Fig. 4.35(c), the output line is pre-charged, and the output from a pixel is compared with the reference voltage

V_{ref} in the column amplifier. The output from the pixel decreases in proportion to the input light intensity, so that PWM is achieved. Using a 3T-APS structure, the sensor achieves a large array format of VGA with a good accuracy of 0.26 mm at a distance of 1.2 m. Refs. [168] and [478] report on the implementation of analog current copiers [489] and comparators to determine the maximum peak quickly. In Ref. [168], analog operation circuits are integrated in a pixel, while in Ref. [478], four sets of analog frame memory are integrated on a chip to reduce the pixel size, and a color QVGA array is realized with a 4T-APS pixel architecture.

In the structured light method, it is important to suppress the ambient light against the structured light, which is incident on the FOV. Combined with a wide dynamic range using a log sensor and modulation technique, Ref. [460] reports a good signal-to-background ratio (SBR) of 36 dB with a dynamic range (DR) of 96 dB*.

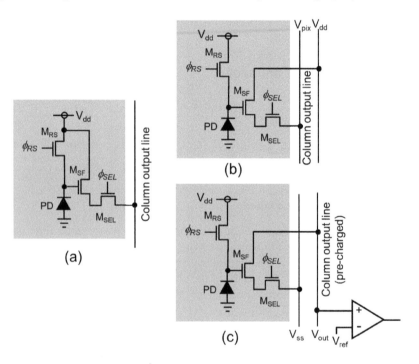

FIGURE 4.35: (a) Conventional 3T-APS structure, modified for (b) normal imaging mode, (c) PWM mode. Illustrated after [487].

---

*In Ref. [460], the definition of SBR and DR are different by a factor of half from the conventional definition. In this book, the conventional definition is adopted.

5

Applications

5.1   Introduction

This chapter is essentially dedicated to introducing numerous and diverse applications of smart CMOS image sensors. The first category of applications considered here relates to information and communication technology (ICT), as applied to transportation and human interface. Introducing imaging functions in these fields can improve their performance, such as communication speed and convenience, for example, in visually aided controls. The second category comprises applications in the field of chemical engineering. By introducing smart CMOS image sensors, chemical properties, such as pH, can be visualized as two-dimensional images. The third category of applications is in the biotechnology field. Here, the CCDs combined with optical microscopes have been widely used. With smart CMOS image sensors, the imaging systems could be made compact with integrated functions, leading to improvements in performance. Furthermore, smart CMOS image sensors may be used to realize characteristics that could not be realized by the CCDs. Finally, the last category introduced in this chapter involves medical applications, which typically require very compact imaging systems, because some of them may be introduced into the human body through swallowing or by means of implant. Smart CMOS image sensors are suitable for these applications because of their small volume, low power consumption, integrated functions, etc. In the above applications, it may be useful to introduce multi-modal sensing and/or stimulation combined with imaging, which cannot be achieved by CCD image sensors.

5.2   Information and communication technology applications

Ever since blue LEDs and white LEDs have emerged, light sources have drastically changed. For example, room lighting, lamps in automobiles, and large outdoor displays nowadays use LEDs. The introduction of LEDs with their fast modulation has produced a new application area, called 'free-space optical communication'. Visible light communication (VLC) is one type of free-space optical communication, involving visible light sources, especially LED-based signals, such as room lights, automobile lumps, traffic signals and outdoor displays [490]. Combined with image sensors, free-space optical communication can be extended to human interfaces because images are visually intuitive for humans. This specific area called 'optical camera communication (OCC)' is a candidate for the standardization of issues in IEEE 802.15.7r1.

In this section, ICT applications of smart CMOS image sensors are introduced. Firstly, the topic of optical wireless communication is introduced, followed by a description of 'optical identification (ID) tag'.

5.2.1 Optical wireless communication

Optical wireless communication or free-space optical communication has advantages over conventional communication that uses optical fibers and RF for the following reasons:

- Setting up an optical wireless system entails only a small investment, which means that such a system could be used for communication between buildings.
- It has the potential of high-speed communication at speeds in Gbps [491].
- It is not significantly affected by interference from other electronic equipment, which is important for use in hospitals, where this may be a critical issue for persons implanted with electronic medical devices such as cardiac pacemakers.
- It is secure because of its narrow diversity.
- It can use wavelength division multiplexing (WDM) with multi-color LEDs.
- It can be simultaneously used with imaging, and this implies the possibility of creating a smart CMOS image sensor with optical communication capability, which is mentioned later.

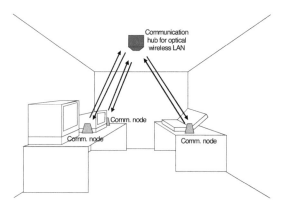

FIGURE 5.1: Example indoor optical wireless communication system. A hub station is installed in a ceiling and several node stations are located near the computers to help them communicate with the hub.

Optical wireless communication systems are classified into three categories based on their usage. The first of these categories involves optical wireless communications for use in outdoors or over long distances beyond 10 m. The main application target here is local-area networks (LANs) across buildings. Optical beams in free space can easily connect two points of a building with a fast data rate. LAN in a factory is another application for this system. The low electro-magnetic interference (EMI) and ease of installation makes optical wireless communication suitable for factories in noisy environments. Many products have been commercialized for these applications.

The second category of applications is in near-distance optical communication. IrDA and its relatives belong in this category [492]. Here, the data rate is not fast, because of the speed of LEDs. This system competes with RF wireless communication systems, such as Bluetooth.

The third category is indoor optical wireless communication, which is similar to the second category, but is intended for use as a LAN, i.e., with a data rate of at least over 10 MHz. Such indoor optical wireless communication systems have been already commercialized but are limited [493]. Figure 5.1 shows an optical wireless communication system for indoor use. The system consists of a hub station installed in a ceiling and multiple node stations located near the computers. It has a 'one-to-many' communication architecture, while the outdoor systems usually have 'one-to-one' communication architecture. This one-to-many architecture suffers from several issues. For example, a node is required to find the hub. To achieve this task, the node is equipped with a mechanism to move the bulky optics with a photo-receiver, which means that the node station is relatively large in size. The size of the hub and node is critical when the system is intended for indoor use.

Recent dramatic improvements in LEDs, including blue and white LEDs, have opened up possibilities for a new type of optical wireless communication called visible light communication (VLC) [490]. The standard IEEE 802.15.7 supports VLC with high data rates of up to 96 Mbps [494]. Many applications using this concept have been proposed and developed; for example, white LED light in a room can be utilized as a transceiver; and LED lamps of automobiles can be used for communication with other automobiles and traffic signals as shown in Fig. 5.2 [495]. High-speed VLCs in indoor use have been developed by using LEDs [491,496,497]. These applications are also related to optical ID tags, which are discussed in Sec. 5.2.2.

In optical wireless communications, it is effective to use 2D detector arrays not only to enhance the receiver efficiency [498] but also to utilize the image itself with data communication. Replacing such 2D detector arrays with a smart CMOS image sensor dedicated to fast optical wireless communication has some advantages, some of which have been proposed and demonstrated by the Nara Institute of Science & Technology (NAIST) [499–502], and by University of California (UC) at Berkeley [503]. The latter worked on communication between small unmanned aerial vehicles [503, 504], meant for outdoor communication over relatively long distances compared with indoor communication.

Another application for optical wireless communication is in automobile transportation systems as shown in Fig. 5.2. Such applications have been developed

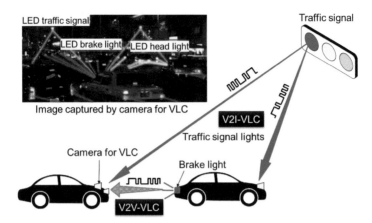

FIGURE 5.2: Conceptual schematic of optical communication for automobile transportation. VLC: visible light communication. V2I: vehicle to infrastructure. V2V: vehicle to vehicle. Illustrated after [495].

by researchers at Keio University, Nagoya Universtiy, and the collaborative group of Toyota and Shizioka University [495, 505–507]. It should be noted that the conventional CMOS image sensors work in accumulation mode and therefore, it is difficult to apply them to high-speed signal detection directly.

In the following sections, firstly, a new scheme of indoor optical wireless LAN is introduced and described in detail, in which an image sensor based photo-receiver is utilized. The photodiode is used in both accumulation and direct photo-current modes. Secondly, CMOS image sensors dedicated for automobile transportation communication system are introduced. These sensors engage specially developed photo-detectors that can detect high-speed modulation optical signals. Two types of pixels are introduced, one for acquiring the image and another for detecting fast modulated optical signals.

5.2.1.1   Optical wireless communication for LAN

In indoor optical wireless LANs, a smart CMOS image sensor is utilized as a photo-receiver as well as a 2D position-sensing device for detecting the positions of communication modules of nodes or the hub. In contrast, in a conventional system, one or more photodiodes are utilized to receive the optical signals.

Figure 5.3 compares the proposed indoor optical wireless LAN system with a conventional system. In the optical wireless LAN, optical signals must be transmitted toward the counterpart accurately to achieve a certain optical input power incident on the photo-detector. Thus the position detection of the communication modules and alignment of the light are significant. In a conventional optical wireless LAN system with automatic node detection, as shown in Fig. 5.3(a), a mechanical scanning system for the photodetection optics is implemented at the hub to search for the

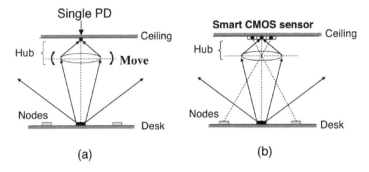

FIGURE 5.3: Optical wireless LAN systems of (a) a conventional indoor optical wireless LAN system and (b) the proposed system using a smart CMOS image sensor.

nodes. However, the scanning system becomes bulky because the diameter of the focusing lens must be large enough to gather sufficient optical power. On the other hand, as shown in Fig. 5.3(b), using a smart image sensor as a photo-receiver is proposed as mentioned before.

The proposed approach includes several excellent features for optical wireless LAN systems. Because an image sensor can capture the surroundings of a communication module, it is easy to specify the positions of other modules by using simple image recognition algorithms without any mechanical components. In addition, image sensors inherently capture multiple optical signals in parallel by means of a huge number of micro-photodiodes. Different modules are detected by independent pixels if the image sensor has sufficient spatial resolution.

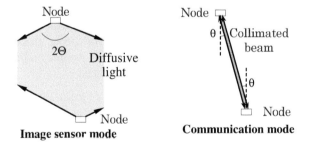

FIGURE 5.4: Two modes in the sensor, (a) image sensor (IS) mode and (b) communication (COM) mode. Adapted from [499] with permission.

The sensor has two functional modes: image sensor (IS) mode and communication (COM) mode. Both the hub and node work in the IS mode, as shown in Fig. 5.4(a). They transmit diffusive light to notify their existence to each other. To cover a large area, where the counterpart possibly exists, diffusive light with a wide radiation angle 2Θ is utilized. Because the optical power detected at the counterpart is generally

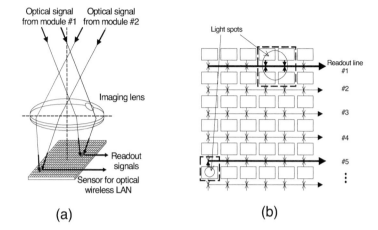

(a)   (b)

FIGURE 5.5: Illustration of the space division multiplexing using a sensor with focused readout, (a) readout of multiple light spots and (b) schematic of the focused readout on the sensor. Adapted from [499] with permission.

very weak, it is effective to detect it in the IS mode. It should be noted that, in a conventional system, the optical transceivers need to scan the other transceivers in the room by swinging bulky optics, which is both time- and energy-consuming.

After the positions are specified in the IS mode, the functional mode of the sensors for both the node and hub are switched to the COM mode. As shown in Fig. 5.4(b), they now emit a narrow collimated beam carrying communication data in the detected direction toward the counterpart. In the COM mode, photocurrents are directly read out without temporal integration from the specific pixels receiving the optical signals. Use of a collimated beam reduces the power consumption at the communication module and receiver circuit area, because the output power at the emitter and the gain of the photo-receiver can be reduced.

Systems using smart CMOS image sensors have a further advantage in that they can employ space division multiplexing (SDM) and thus increase the communication bandwidth. Figure 5.5 illustrates SDM in a system. When optical signals from the different communication modules are received by independent pixels and read out to separate readout lines, concurrent data acquisition from multiple modules can be achieved. Consequently, the total bandwidth of the downlink at the hub can be increased in proportion to the number of readout lines.

To read out the photo-current in the COM mode, a 'focused readout' is utilized. As shown in Fig. 5.5(b), a light signal from a communication module is received by one or a few pixels, because the focused light spot has a finite size. The amplified photocurrents from the pixels receiving the optical signals are summed to the same readout line prepared for the COM mode, and hence, the signal level is not reduced.

Implementation of a smart CMOS sensor Figure 5.6 shows the pixel circuit, which consists of a 3T-APS, transimpedance amplifier (TIA), digital circuitry for mode switching, and latch memory. To select either COM or IS mode, a HIGH or LOW signal is written in the latch memory. The output from the TIA is converted to a current signal to sum the signal from the neighboring pixels. As shown in Fig. 5.6, each pixel has two data–output lines, one on the left and another on the right, for focused readout as mentioned above. The sum of the current signal flows is input into a column line and converted with the TIA and amplified with the main amplifier.

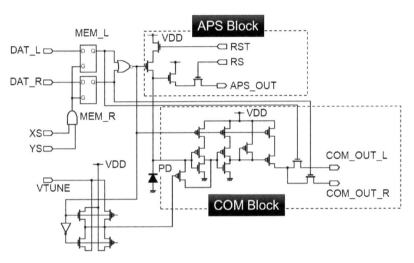

FIGURE 5.6: Pixel circuits of a smart CMOS image sensor for an indoor optical wireless LAN. It consists of an APS block for image sensing and a COM block for high-speed detection of optical signal. Illustrated after [499].

A block diagram of a sensor used for indoor wireless LAN is shown in Fig. 5.7 (a). The sensor has one image output and four data outputs. The image is read out through a sample and hold (S/H) circuit. The chip has four independent data output channels for the COM mode. It is fabricated in standard 0.35-μm CMOS technology. A micro-photograph of the fabricated chip is shown in Fig. 5.7(b).

Figure 5.8 shows experimental results for imaging and communication using the fabricated sensor. Its photosensitivity at 830 nm is 70 V/(s·mW). The received waveform is shown in Fig. 5.8. The eye was open with 50 Mbps at 650 nm and 30 Mbps at 830 nm. In this sensor, an intense laser light is incident for fast communication on a few pixels, while the other pixels operate in image sensor mode. The laser beam incident on the sensor produces a large number of photo-carriers, which travel a long distance, according to the wavelength. Some of the photo-generated carriers enter the photodiodes and affect the image. Figure 5.9 shows experimental results of effective diffusion length measured in this sensor for two wavelengths. As expected, the longer wavelength has a longer diffusion length, as discussed in Sec. 2.2.2.

Further study is required to increase the communication data rate. Experimental results using 0.8-μm BiCMOS technology show that a data rate of 400 Mpbs/channel can be obtained. Also, a system introducing the wavelength division multiplexing (WDM) has been developed [501, 502], that can increase the data rate.

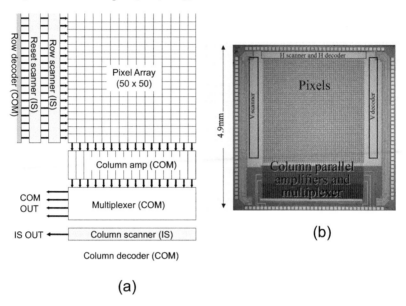

FIGURE 5.7: A smart CMOS image sensor for an indoor optical wireless LAN. (a) Block diagram of the sensor, and (b) Chip photo. with permission [500].

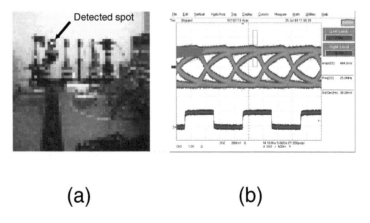

FIGURE 5.8: Experimental results of the smart CMOS image sensor: (a) Captured image, and (b) 30-Mbps eye pattern for a wavelength of 830 nm. Adapted from [500] with permission.

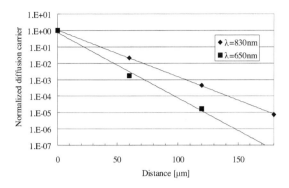

FIGURE 5.9: Measured amount of diffusion carriers on the pixels. Adapted from [500] with permission.

5.2.1.2 Optical wireless communication for automobile transportation system

An optical wireless communication can be utilized for an automobile transportation system as shown in Fig. 5.10. The smart CMOS image sensor installed in the automobile can detect and modulate signal from a stoplight and demodulate data from the modulated signal. Several methods to detect modulated signals have been developed as described in Sec. 4.5 of Chapter 4.

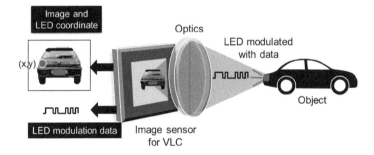

FIGURE 5.10: Visible light communication (VLC) using a smart CMOS image sensor. The light from an LED is modulated to carry the data, and the smart image sensor can detect the modulation data and capture the image in one chip.

Implementation of a smart CMOS sensor The smart CMOS image sensor has been developed to detect modulated signals, as well as take images in a conventional manner [495, 506, 507]. It should be noted that the modulated signal is so fast that it is difficult to detect such high-speed signals by using conventional image sensors. In this sensor, the pixel for acquiring image and the pixel for fast optical communication are placed in neighboring column as shown in Fig. 5.11 [507].

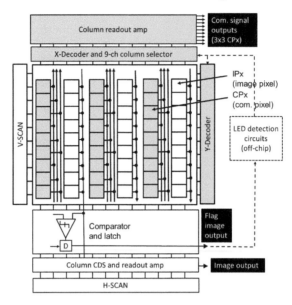

FIGURE 5.11: Chip block diagram. The white area shows the imaging functions and the gray area shows the optical communication functions. Illustrated after [507].

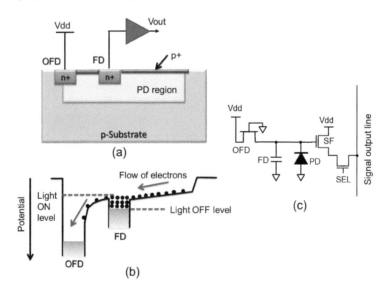

FIGURE 5.12: Pixel for optical communication: (a) cross-section of the pixel, (b) potential profile, and (c) pixel schematic. FD: floating diffusion. OFD: over-flow drain. Illustrated after [507].

The first step is to find out the modulated optical signals in acquired image. By using a comparator and a latch placed in each column, a flag image is output as one-bit image, and is processed to detect the modulated optical signals in the image. The next step is to assign ROIs (3 x 3 pixels) for the modulated optical signal and activate the communication pixels (3 x 3 pixels). To achieve the detection of high-speed modulated optical signals, a pixel that is built in an internal electric field that makes the photo-generated electrons drift fast is shown in Fig. 5.12. Over-flown electrons are swept out to the overflow drain (OFD) .

5.2.2 Optical ID tag

The recent spread of large outdoor LED displays, white LED lighting, etc. has prompted research and development into applying LEDs to transceivers in spatially optical communication systems. Visible light communication is one such system [490, 508]. This application is described in Sec. 5.2.1.

An alternative application is an optical ID tag, which uses an LED as a high-speed modulator to send the ID signals. Applying optical ID tags combined with image sensors to augmented reality (AR) technologies has been proposed and developed as shown in Fig. 5.13. However, conventional CMOS image sensors can only detect the LED signals with low data rate of approximately 10 bps [509].

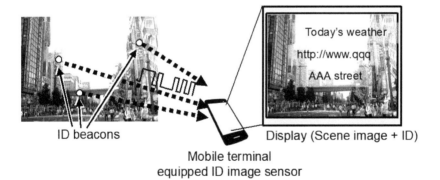

FIGURE 5.13: Concept of optical ID tag. LEDs on a large LED display are used as optical ID tags. ID data is displayed on the image taken by the user.

There are two approaches to overcome the data rate limitation. One is to use conventional image sensors with WDM or multi-color LEDs. This method is used in "Picapicamera^{TM}," or "Picalico^{TM}" [510–512], and "FlowSign Light^{TM}" [513,514]. In another method, conventional CMOS image sensors are used to introduce multiple sampling as in "LinkRay^{TM}" [515,516] and in [517].

The second approach is to use smart CMOS image sensors that are dedicated for detecting the LED modulation signals, such as "ID cam" [168,518,519], "OptNavi" [520–522], and others [523–525].

5.2.2.1   Optical ID tags with conventional CMOS image sensors

These systems have been developed for optical ID tag systems using conventional CMOS image sensors. To apply conventional CMOS image sensors to detect optical ID signals, LinkRay uses rolling shutter mechanism in conventional CMOS image sensors as shown in Fig. 5.14 [516]. Danakis *et al.* also demonstrated a similar method [517]. Date rate with a few kbps can be achieved by the method. In Picapicamera, the RGB color modulation method was used and a data rate of about 10 bps was achieved for data transmission [511, 512]. Although the date rate is not so high in Picapicamera, it is assumed by combining Picapicamera with the internet, rich content can be used. FlowSign Light also uses color modulation of LEDs and achieves a bit rate of 10 bps, similar to Picapicamera [514]. Although these methods are suitable for smart phone application, they have a limitation of data rate.

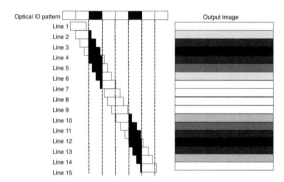

FIGURE 5.14: Optical ID tags with conventional CMOS image sensors. Illustrated after [516].

5.2.2.2   Optical ID tags with smart CMOS image sensors

The ID cam, proposed by Matsushita *et al.* at Sony, uses an active LED source as an optical beacon with a dedicated smart CMOS image sensor to decode the transmitted ID data [168]. The smart sensor used in the ID cam is described later.

OptNavi has been proposed by the author's group at NAIST and is designed for use with mobile phones, as well the ID cam [520, 526, 527]. One typical application of these systems, LEDs on a large LED display, can be used as an optical beacon, as shown in Fig. 5.13. In this figure, the LED displays send their own ID data. When a user takes an image using a smart CMOS image sensor that can decode these IDs, the decoded data is superimposed on the user interface, as shown in Fig. 5.13. The user can easily get information about the content on the display as AR.

An alternative application is a visual remote controller for electronic appliances connected together in a network as IoT (Internet of Things). The OptNavi system has been proposed for use in a man–machine interface to visually control networked appliances. In this system, a mobile phone with a custom image sensor is used as an

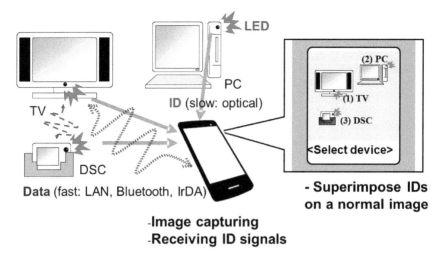

FIGURE 5.15: Concept of OptNavi. A dedicated smart CMOS image sensor can detect and decode optical IDs from home electronic appliances. The decoded results are superimposed on the image taken by the sensor.

interface. Smart phones are equipped with a large display, a digital still camera, IrDA [528], Bluetooth[TM] [529], etc. In the OptNavi system, home network appliances, such as a DVD recorder, TV, and PC, are equipped with LEDs that transmit ID signals at about 500 Hz. The image sensor can receive IDs with a high-speed readout of multiple ROIs. The received IDs are displayed with a superimposed background image captured by the sensor, as shown in Fig. 5.15. With the OptNavi system, we can control such appliances by visually confirming them on a mobile phone display.

Because a conventional image sensor captures images at a video frame rate of 30 fps, it cannot receive optical ID tag signals at a kHz rate. Thus, a custom image sensor with a function for receiving the optical ID signals is needed. Several smart CMOS image sensors dedicated for optical ID tags have been proposed and demonstrated [168, 478, 523–527, 530], and are summarized in Table 5.1. These sensors receive ID signals with a high-speed frame rate. S. Yoshimura *et al.* and T. Sugiyama *et al.* from the same group have demonstrated image sensors that operate with high-speed readouts for all pixels with the same pixel architecture as a conventional CMOS image sensor [168, 478]. These sensors are suitable for high resolution. Y. Oike *et al.* has demonstrated a sensor that receives ID signals by analog circuits in a pixel, capturing images at a conventional frame rate [524]. This sensor can receive ID signals with high accuracy. J. Deguchi *et al.* has demonstrated a HD CMOS image sensor that receives ID signals with scanning ROI many times in row direction [525]. By introducing this multiple row re-scan method, the architecture of the pixel is almost the same as a conventional CMOS image sensor and hence, the pixel size can be shrunk to $2.2 \times 2.2 \mu m^2$.

TABLE 5.1
Specifications of smart CMOS image sensors for optical ID tags

Affiliation	Sony	Univ. Tokyo	NAIST	Toshiba
Reference	[478]	[523]	[526]	[525]
ID detection	High-speed readout of all pixels	In-pixel ID receiver	Readout of multiple ROIs	Multiple row rescan
Technology	0.35μm	0.35μm	0.35μm	0.13μm
Pixel size	$11.2 \times 11.2 \ \mu$m^2	$26 \times 26 \ \mu$m^2	$7.5 \times 7.5 \ \mu$m^2	$2.2 \times 2.2\mu$m^2
Pixel count	320×240	128×128	320×240	1720×832
Frame rate for ID detection	14.2 kfps	160 k fps	1.2 kfps	30fps
Bit rate	120 bps	4.85 kbps	4 kbps	1.92 kbps
Power consumption	82mW (@3.3 kfps, 3.3 V)	682mW (@4.2 V)	3.6mW (@ 3.3 V, w/o ADC, TG)	472mW (w/ ADC)

Implementation of smart CMOS image sensors for optical ID tags   High-speed readout of all pixels may cause large power consumption. The NAIST group has proposed an image sensor dedicated for optical ID tags to realize high-speed readout at low power consumption with a simple pixel circuit [527]. In the readout scheme, the sensor is operated for capturing normal images at a conventional video frame rate, while simultaneously capturing multiple ROIs which receive ID signals at a high-speed frame rate. To locate ROIs with the sensor, an optical pilot signal which blinks at a slow rate of about 10 Hz is introduced; the sensor can easily recognize it with a frame difference method, as shown in Fig. 5.16.

Figure 5.17 shows a block diagram of the sensor. The sensor is operated with an ID map table, which is a 1-bit memory array with the ID positions memorized for reading out the ROIs at high-speed. Furthermore, the power supply to the pixels outside the ROIs is cut off. The pixel circuit is simple; it has only one additional transistor for column reset compared with a conventional 3T-APS, as shown in Fig. 5.17(b). It should be noted that an additional transistor must be inserted between the reset line and the gate of the reset transistor, as described in Sec. 2.6.1. This transistor is used for XY-address reset or random access as described in Sec. 2.6.1 to read out the pixels only in the ROIs.

Figure 5.17(c) shows a micro-photograph of the sensor. Figure 5.18(a) shows a captured normal image at 30 fps and Figs. 5.18(b) and (c) show experimental results for the ID detection. The ID signals are transmitted by a differential 8-bit code modulated at 500 Hz from three LED modules. 36-frame ROI images per ID, which consist of 5×5 pixels, are captured for detecting the ID signals, while one frame of a whole image is captured, as shown in Fig. 5.18(c). The patterns of the ROI images successfully demonstrate the detection of each ID.

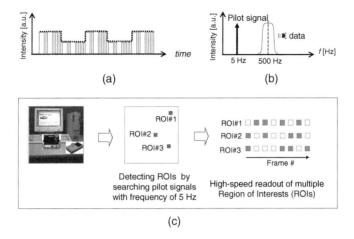

(a)   (b)

(c)

FIGURE 5.16: Method of detecting ID positions using a slow pilot signal: (a) waveforms of pilot and ID signals, (b) power spectrum of pilot and ID signals, and (c) procedure to obtain the ID signal using the pilot signal [527].

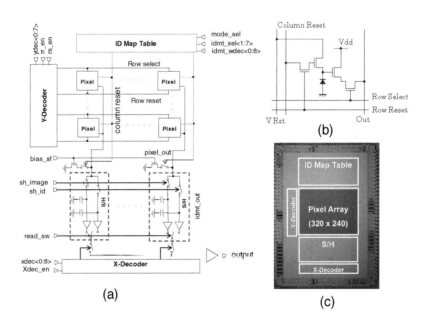

(a)   (c)

FIGURE 5.17: (a) Block diagram of a smart CMOS image sensor for fast readout of multiple ROIs. (b) Pixel circuits. (c) Chip photograph. Adapted from [526] with permission.

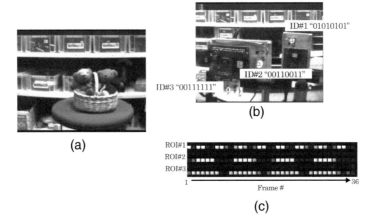

FIGURE 5.18: Images taken by a smart CMOS image sensor for optical ID tags: (a) Normal image, (b) ID detection and (c) 36-frame ROI images/ID (in one frame for a normal image). Adapted from [526] with permission.

5.3 Chemical applications

CMOS image sensors for chemical applications are mainly meant for imaging chemical parameters, such as optical activity (chirality) and pH. In this section, firstly, smart CMOS image sensors for detecting optical activity are introduced. The basic structure of the sensor is covered in Sec. 3.4.5 of Chapter 3. Next, pH imaging sensors are described. The pH imaging sensors can be applied for the detection of chemicals, such as acetylcholine by changing the receptive film on the sensing area.

5.3.1 Optical activity imaging

In Sec. 3.4.5 of Chapter 3, polarimetric image sensors are mentioned. In this section, optical activity imaging is described as one of the chemical industry applications of the polarimetric image sensors. In chemical synthesis, chirality is crucial. For example, optical activity is the same as optical rotation. When a linearly polarized light enters into a solution with optical activity, the polarization angle is rotated by $\alpha°$ as shown in Fig. 5.19.

The specific optical rotation angle $[\alpha]$ is defined in the following conditions:

- The density of solution: 1 [g/ml]
- The temperature of solution: 20 [°C]
- The solvent: water
- Wavelength of light used: Na D-line (589 [nm])

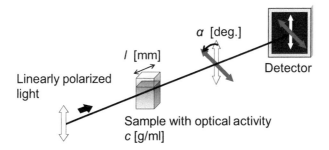

FIGURE 5.19: Optical activity. When linearly polarized light enters into a sample with optical activity, the polarization angle is rotated by $\alpha°$ after the sample. ℓ : sample length [mm] along the light path; c: density of the sample [g/ml].

Under these conditions, $[\alpha]$ is expressed as

$$[\alpha] = \frac{100\alpha}{\ell c}, \tag{5.1}$$

where ℓ is the sample length in [mm] and c is the density of the sample in [g/ml].

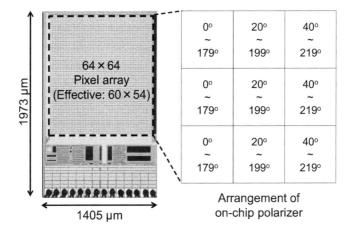

FIGURE 5.20: Chip photo of polarization detection image sensor. Adapted from [531] with permission.

Conventionally, $[\alpha]$ is measured with a polarimeter. In a polarimeter, a polarizer is rotated to find out the optical rotation angle and thus, it is bulky and takes time to measure. By using a polarization detection image sensor, mentioned in Sec. 3.4.5 of Chapter 3, $[\alpha]$ can be obtained at a particular time. While in a polarimeter, the polarization angle is varied in time, in a polarization detection image sensor, it is varied in a space as shown in Fig. 3.43(b). The polarization detection image sensor can monitor the optical activity in chemical reaction process in real time. This feature can provide real time feedback of chemical reactions.

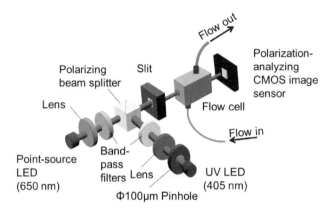

FIGURE 5.21: Experimental system of optical rotation angle measurement using a polarization detection image sensor. The absorbance is simultaneously measured. The red LED is used for polarization angle measurement, while the ultra violet (UV) LED is used for optical absorbance measurement.

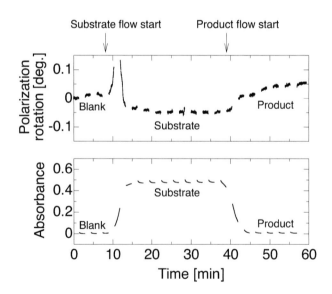

FIGURE 5.22: Experimental results of optical rotation angle and optical absorbance in a chemical reaction by using a polarization detection image sensor. Adapted from [531] with permission.

By using the sensor shown in Fig. 5.20, a real chemical reaction can be monitored [531]. In this sensor, pixels could be either with a metal wire grid or without a metal wire grid, and by using a pixel without the wire grid, the optical absorbance can be measured with $[\alpha]$ simultaneously. Such a multi-modal measurement is also an advantage of smart CMOS image sensors. Figure 5.21 illustrates the experimental setup of polarization detection system and the optical absorbance by using a polarization detection image sensor. The chiral chemical solution is flowed into the optical cell. Figure 5.22 shows typical experimental results obtained by using the polarization detection image sensor. The data values of polarization rotation angle and the absorbance are plotted with time in the figure. When the product flow starts, $[\alpha]$ begins to change, while the optical absorbance is changed. These results clearly demonstrate the effectiveness of polarization detection image sensors for real-time monitoring of optical activity.

5.3.2 pH imaging sensor

In this section, applications of smart CMOS image sensors to pH or proton imaging are introduced. Measuring the pH value is important for quality monitoring of water, food and soil. Conventional pH meters measure pH at a point, while pH image sensor can measure a 2D pH distribution and monitor it in real time. Two types of smart image sensors are introduced here: one is a light-addressable potentiometric sensor, (LAPS); and the other is an ion sensitive field effect transistor (ISFET)-based pH imaging sensor.

5.3.2.1 LAPS: light-addressable potentiometric sensor

LAPS was first proposed and demonstrated by D.G. Hafeman *et al.* [532], and was developed by Yoshinobu *et al.* [533–536]. The structure of LAPS is based on the electrolyte-insulator-semiconductor (EIS) structure as shown in Fig. 5.23.

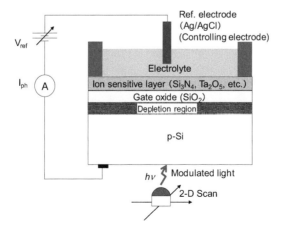

FIGURE 5.23: LAPS: light-addressable potentiometric sensor.

The change of pH in the electrolyte modulates the surface potential of the semiconductor. A modulated photo-current flows by illuminating the backside of the semiconductor (in this case, Si) with modulated light. By measuring the photocurrent, pH can be estimated [536]. By 2D scanning with the modulated light source, a 2D map of pH can be obtained.

5.3.2.2 ISFET-based pH sensor with accumulation mode

ISFET is one variation of EIS capacitance structure mentioned above [537–539]. Just as MOSFET is a MOS capacitor structure with source and drain contacts, ISFET is an EIS capacitor structure with source and drain contacts as shown in Fig. 5.24. In ISFET, the threshold voltage is modulated by the change of amount of charges in the solution or change of pH, ΔV_{pH}. ΔV_{pH} follows Nernst equation as

$$\Delta V_{pH} = \frac{RT}{F} ln \frac{a_{H^+}^I}{a_{H^+}^{II}} \tag{5.2}$$

$$\simeq 60mV/\Delta pH, \tag{5.3}$$

where R, F, and T are gas constant, Faraday constant, and absolute temperature, respectively. $a_{H^+}^i$ is an activity of proton H^+ in i region (i=I, II). The last equation shows the limitation of the sensitivity for pH detection system.

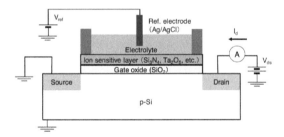

FIGURE 5.24: Structure of ISFET.

To extend the limitation of Nernst equation in Eq. 5.3, the accumulation mode in image sensors is introduced [540, 541] as shown in Fig. 5.25(a). The structure is similar to that of an ISFET; however, on either side of the ion sensitive layer, a gate electrode is located. One is called the input gate, while the other is referred to as the output gate. In addition, the source region and the drain region in the ISFET are used as an input diode and a floating diffusion, respectively. The drain region is connected to a source follower and is similar to a 4T-APS. The operation principle is as shown in Fig. 5.25(b) and described in Ref. [541]. The potential in an ISFET is produced by the amount of protons or positive charges in the solution. Furthermore, in this sensor, the potential in the semiconductor surface is produced according to the amount of pH. In step (1) in Fig. 5.25(b), the potential is produced by pH. In step (2), after the input diode injects the electrons, the input gate is opened, the electrons flow into the potential region, and fill it. In step (3), the input gate is closed, so that the potential region is isolated. In step (4), the output gate is opened. Sufficient gate voltage is applied so as to flow all the electrons in the potential region into the FD. Finally, the status is returned to the initial step, i.e., (1). On repeating this process, the amount of electrons in the potential is accumulated in the FD, so that the total SNR increase as $\sqrt{N}$ where N is the number of repetitions. This method can overcome the limitation of Nernst equation.

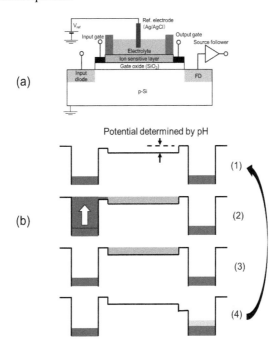

FIGURE 5.25: Structure and operation sequence of pH image sensor using accumulation mode. Illustrated after [541].

5.4 Bioscience and Biotechnology applications

In this section, applications of smart CMOS image sensors to bioscience and biotechnology are introduced. Fluorescence detection, a widely used measurement in such applications and conventionally performed by a CCD camera installed in an optical microscope system, has been identified as an application that could be efficiently performed by smart CMOS image sensors. Introducing smart CMOS image sensors into bioscience and biotechnology would leverage the benefits of integrated functions and miniaturization. Three types of applications of smart CMOS image sensors are introduced for these applications, that is, attachment type, on-chip type, and implantation type, which are shown in Fig. 5.26.

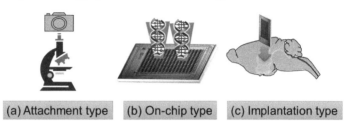

| (a) Attachment type | (b) On-chip type | (c) Implantation type |

FIGURE 5.26: Three types of bio applications using smart CMOS image sensors: (a) attachment type, (b) on-chip type, and (c) implantation type.

As an example of the attachment type (Fig. 5.26(a)), smart CMOS image sensors with high-speed response can be applied to detect the lifetime of fluorescence, which is called fluorescence lifetime imaging microscopy (FLIM). Similar to this example, smart CMOS image sensors used with conventional optical equipment, such as an optical microscope, are suitable for bioscience and biotechnology applications.

The second type is the on-chip detection type [371, 542–546]. On-chip type means that a specimen can be placed directly on the chip's surface and measured as shown in Fig. 5.26 (b). Such a configuration would make it easy to access a specimen directly so that parameters, such as fluorescence, potential, pH [545], and electrochemical parameters can be measured. Integration functions are important features of a smart CMOS image sensor. These functions realize not only high SNR measurements but also functional measurements. For example, electrical simulation can be integrated into a sensor so that fluorescence can be caused by cell stimulation; this is an on-chip electro-physiological measurement.

Through miniaturization, the total size of the sensing system can be reduced. Such miniaturization makes it possible to implant the sensing system. Also, on-field measurements are possible because the reduced size of the total system enhances its mobility. This is the third type, i.e., the implantation type shown in Fig. 5.26(c). This example involves implantation of a smart CMOS sensor in a mouse brain. In this case, the sensor is made small enough to be inserted into a mouse brain. Inside the brain, the sensor can detect fluorescence as well as stimulate neurons around it.

This section introduces three example systems that exhibit the advantages described above. The first is an on-chip fluorescence detection with a CMOS image sensor, in which each pixel has stacked PDs to discriminate excitation light and fluorescence. The second is a multi-modal image sensor, which takes electrostatic images or electrochemical images as well as performing optical imaging [547–549]. The third is an *in vivo** CMOS image sensor [550–553].

In vivo means "within a living organism." Similarly, *in vitro* means "in an artificial environment outside a living organism."

5.4.1 Attachment type

5.4.1.1 FLIM

As a typical example of the attachment type, smart CMOS image sensors for FLIM are considered. Figure 5.27 shows the basic structure for FLIM measurement.

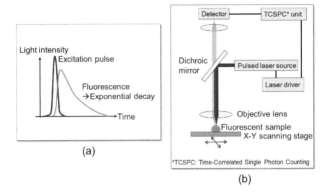

FIGURE 5.27: Fundamental of FLIM (a) Response of excitation pulse and fluorescence decay. (b) Basic experimental setup.

The excitation light pulse only emits very short light pulses. After the fluorescence is triggered by the excitation light pulse, it starts to decay and finally diminishes. It is very important to measure this fluorescence decay time for the molecular information of neural cells, which helps us understand their corresponding molecular configurations. As the decay time is normally very short, i.e., of the order of nanoseconds, high-speed imaging with nanosecond response time is required. However, we know that the response time in conventional image sensor is milliseconds, which means that it is very difficult to measure such a short decay time using conventional CMOS image sensors. A conventional FLIM microscope usually uses one detector, which is generally an avalanche photodiode (APD) with a very fast response time. The time-correlated single photon counting technique is employed to measure the correlation between the measured pulse and the fluorescence decay as well as the distribution of the fluorescence lifetime. Obviously, it is a very time-consuming process.

If we use a CMOS image sensor instead of just one detector, we can obtain the distribution of fluorescence lifetime in real time. One major issue is that the response time of the conventional CMOS image sensor is very slow. So how can we enhance the response speed of CMOS image sensor? The main speed limitation for a 3T-APS is the response time of the photodiode, which ultimately limits the overall readout speed. For a 4T-APS, the transition time for the charge to transfer from the photodiode capacitance to the FD can result in an extra delay for the signal readout, and determines the speed of the CMOS image sensor. To enhance the speed of CMOS image sensor, one solution is to use a fast detector, such as an APD. Another solution is to introduce drift mechanism during the charge transfer.

It is difficult to achieve uniform sensitivity over a number of APD pixels, because the gain of the APD is an analog value. As a SPAD sensor produces a digital pulse, SPAD sensor array is realistic as described in Sec. 3.2.2.3 of Chapter 3. Alternatively, VAPDs described in Sec. 2.3.4.2 of Chapter 2 may be used because of its digital outputs. By using an array of SPAD sensors, we can obtain a very fast response time, and the FLIM signal can be measured through the delay of these short pulses. To apply a SPAD sensor array to FLIM measurement, two methods have been developed, time-correlated single photon counting (TCSP) method and gate window (GW) method. Figure 5.28 shows the timing charts of the two methods [554]. The TCSP method is used in conventional FLIM.

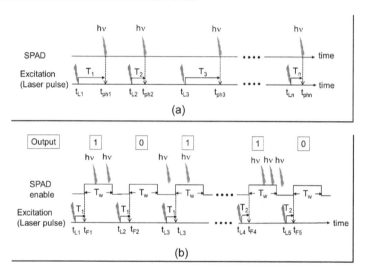

FIGURE 5.28: SPAD sensor array for FLIM measurement. (a) TCSP (time correlated single photon counting) and (b) GW (gate window) methods. Illustrated after [554].

The circuits for TCSP are integrated in one pixel and hence, the pixel size becomes large. In addition, only single photon must be incident, and thus the light intensity should be weak. The GW method is used to turn on and off the SPAD sensor. If the photons hit on the SPAD during the turning on, the output is one, and otherwise zero. To change the width of the window, the decay time can be estimated.

Besides the SPAD, another method is available, which involves introducing a lock-in pixel with drift mechanism during the transfer action as described in Sec. 3.2.1.4 of Chapter 3. Here we show the top view and the cross-sections of the photodiode as shown in Figure 4.14 in Sec. 4.3.4 of Chapter 4. In each pixel, we can use four electrodes to apply different voltages to control the potential profile. By changing the potential inside the photodiode, we can either prohibit or enhance the charge flow using the drift mechanism so as to drastically increase the charge transfer time to the nanosecond regime. By measuring the charge contents at different time instances, we can easily calculate the delay of the fluorescence.

By using the lock-in pixel with the two-FD pixels, which is shown in Fig. 4.14(a) of Sec. 4.3.4 of Chapter 4, we can measure the ultra-fast fluorescence decay and realize FLIM using a CMOS image sensor.

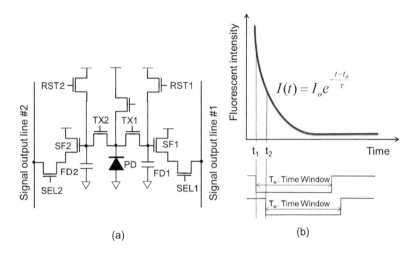

FIGURE 5.29: FLIM measurement by using modulation-based smart pixel: (a) pixel circuit and (b) the measurement principle to obtain the lifetime τ by using the pixel circuit. Illustrated after [555].

In the sensor as shown in Fig. 5.29, the GW method is applied to achieve FLIM measurement [555, 556]. By using two GWs, the lifetime τ can be obtained under the assumption of exponential decay curve $I(t) = I_0 \exp(-\frac{t-t_0}{\tau})$ as shown in Fig. 5.29(b), where $I(t)$ is the fluorescence intensity at time t, I_0 is the initial fluorescence intensity at the initial time t_0, and τ is the fluorescence lifetime. The fluorescence lifetime τ is expressed as

$$\tau = \frac{t_2 - t_1}{\log \frac{V_1}{V_2}}, \tag{5.4}$$

where $V(1)$ and $V(2)$ are the output signal values accumulated during the time window, T_w started at time t_1 and t_2, respectively.

5.4.2 On-chip type

5.4.2.1 On-chip fluorescence imaging

In this section, on-chip fluorescence imaging by using smart CMOS image sensors is introduced. To measure fluorescence, a solution well is required to maintain biological samples such as DNA, virus, etc. Figure 5.30 shows the basic configuration of on-chip fluorescence detection by using solution wells or chambers where bio-markers are contained in a solution such as PBS (phosphate buffered saline).

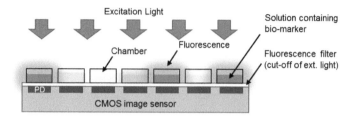

FIGURE 5.30: On-chip fluorescence detection.

5.4.2.1.1 Digital ELISA To demonstrate the on-chip fluorescence detection, digital ELISA [557] combined with CMOS image sensors is described in detail. ELISA, an abbreviation of Enzyme-Linked Immuno-Sorbent Assay, is widely used to examine antigens such as influenza virus, by using antigen-antibody reactions. Figure 5.31 shows the principle of ELISA and the digital ELISA.

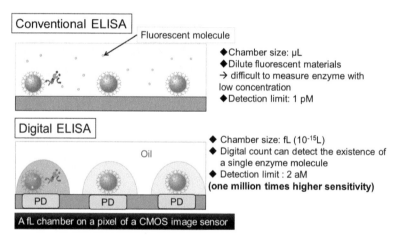

FIGURE 5.31: Principle of digital ELISA.

In ELISA, a series of antigen-antibody reactions are carried out to capture specific bio-markers or antigens and link the enzymes corresponding to the bio-markers. Then, an enzyme-catalyzed fluorescent reaction proceeds and the fluorescence intensity is increased as a result of this reaction. The antibodies are fixed on the surface of beads in a chamber as shown in Fig. 5.31. The analog intensity of fluorescence in a well is proportional to the density of the antigen or virus. The volume of the reaction chambers is of micro-liter scale in a conventional ELISA.

On the contrary, digital ELISA consists of an array of femto-liter-scale micro-chambers [557]. The concentration of the enzyme (corresponding to the target bio-marker) is determined by counting the number of micro-chambers with and without fluorescence as shown in Fig. 5.31. By counting the number of bright chambers, the density of bio-markers can be obtained with a high sensitivity compared with the conventional ELISA. This is the basic principle of digital ELISA. Fluorescence optical microscopes are used to count the number of bright chambers. If the femto-liter-scale micro-chamber array is placed on an image sensor as shown in Fig. 5.31, each pixel can report the existence of a bio-marker so that the virus density can be known with high sensitivity [366, 558–561].

Smart CMOS image sensor with stacked PDs for fluorescence detection   One of the issues in detecting fluorescence with digital ELISA combined with a CMOS image sensor is how to reduce the effect of excitation light, as the intensity of the same is much larger than that of fluorescence light. As scattered excitation light enters the filter in any direction and the interference filter can work only for normal incident light, some portion of the excitation light enters into the CMOS image sensor. To deal with this issue, a CMOS image sensor with stacked PDs has been introduced [366, 559] as shown in Fig. 5.32, which has similar structure in Sec. 3.3.5 of Chapter 3.

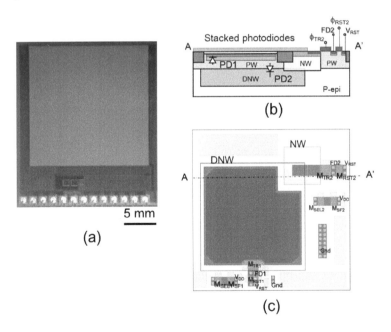

FIGURE 5.32: CMOS image sensor with stacked PD: (a) photograph of the fabricated stacked PD CMOS image sensor, (b) cross-section, and (c) layout of the pixel. Adapted from [366] with permission.

Since the position of pn junction is different between two PDs, the wavelength at peak sensitivity is different. This feature of the stacked PD CMOS image sensor indicates that it can discriminate fluorescence from excitation light. Figure 5.34 shows the digital ELISA system by using CMOS image sensor with stacked PDs. A micro-fluidic device made of PDMS (polydimethylsiloxane) is used to form a micro-chamber array. The detailed structure is described in [366].

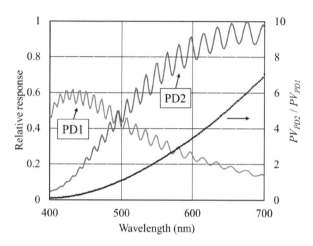

FIGURE 5.33: Sensitivity of the Stacked PDs. Adapted from [366] with permission.

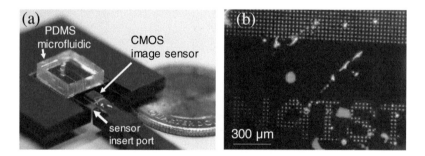

FIGURE 5.34: Digital ELISA system using the CMOS image sensor with stacked PD [366]. (a) photograph of a PDMS microfluidic device set on a micro-chamber array, and (b) droplets of aqueous fluorescent solution on the micro-chamber array taken by the fluorescence microscope. The droplets were formed at positions where the micro-reaction chamber was designed.

5.4.2.2 On-chip multi-modal functions

In this section, smart CMOS image sensors with multi-modal functions are introduced. Multi-modal functions are particularly effective for biotechnology. For example, DNA identification is more accurate and correct if it is combined with an optical image with other physical values such as an electric potential image. Two examples are introduced here, optical-potential multiple imaging [547] and optical-electrochemical imaging [549].

5.4.2.2.1 Optical and potential imaging

Design of sensor A micro-photograph of a fabricated sensor is shown in Fig. 5.35. The sensor has a QCIF (176 × 144) pixel array consisting of alternatively aligned 88 × 144 optical sensing pixels and 88 × 144 potential sensing pixels. The size of the pixels is 7.5 μm × 7.5 μm. The sensor was fabricated in 0.35-μm 2-poly, 4-metal standard CMOS technology.

FIGURE 5.35: Micro-photographs of a fabricated smart CMOS image sensor for optical and potential dual imaging. Adapted from [547] with permission.

Figure 5.36 shows circuits for a light-sensing pixel, a potential-sensing pixel, and the column unit. A potential-sensing pixel consists of a sensing electrode, a source follower amplifier, and a select transistor. The sensing electrode is designed with a top metal layer and is covered with a passivation layer of a standard CMOS process, that is silicon nitride (Si_3N_4). The sensing electrode is capacitively coupled with the potential at the chip surface. While using the capacitive coupling measurement method, no current flows from the image sensor and perturbation caused by the measurement is smaller than that caused by a conductive coupling sensor system, such as a multiple electrode array.

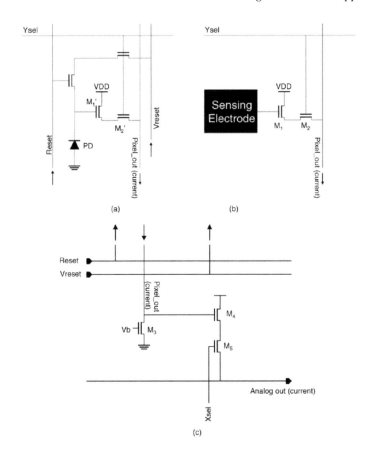

FIGURE 5.36: Circuits of a smart CMOS image sensor for optical and potential dual imaging for (a) a light-sensing pixel, (b) a potential-sensing pixel, and (c) a column unit. Adapted from [547] with permission.

Experimental results   Figure 5.37 shows an as-captured (optical and potential) image and reconstructed images. The sensor is molded from silicone rubber and only a part of the sensor array is exposed to the saline solution in which the sensor is immersed. A voltage source controls the potential of the saline solution via an Ag/AgCl electrode dipped in the solution. As shown in Fig. 5.37, the captured image is complicated because the optical and potential images are superimposed in the image. However, the data can be divided into two different images. The dust and scratches observed in the micro-photograph of Fig. 5.37(a) and a shadow of the Ag/AgCl electrode are clearly observed in the optical image as shown in Fig. 5.37(b). On the other hand, the potential at the exposed region shows a clear contrast to the covered region in the potential image in Fig. 5.37(c). As described later, the potential-sensing pixel shows a pixel-dependent offset caused by trapped

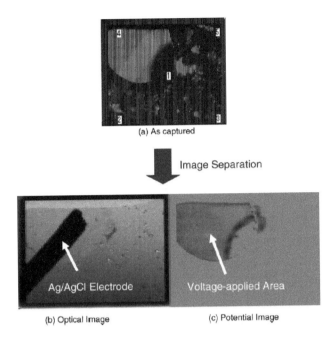

(a) As captured

Image Separation

Ag/AgCl Electrode   Voltage-applied Area

(b) Optical Image   (c) Potential Image

FIGURE 5.37: Images taken by the sensor: (a) as-captured image, (b) reconstructed optical images from image (a), and (c) reconstructed potential image from image (a). Adapted from [547] with permission.

charge in the sensing electrode. However, the offsets are effectively canceled in the image reconstruction process. In the case of a sine wave between 0 and 3.3 V (0.2 Hz) applied to the saline solution, the potential change in the solution can be clearly observed in the region that is exposed to the solution. No crosstalk signal was observed in either the covered potential-sensing pixels or the optical image. Simultaneous optical and potential imaging was successfully achieved.

Figure 5.38 shows the potential imaging results using conductive gel probes. Two conductive gel probes were placed on the sensor and the voltages of the gels were controlled independently. The images clearly show the voltages applied to the gel spots. The difference in applied voltage was captured with good contrast. The sensor is capable of taking not only still images but also moving the images over 1–10 fps. The resolution is currently smaller than 6.6 mV, which is sufficient to detect the DNA hybridization [562]. It is expected that the potential resolution will be improved to 10-μV level. Since the present sensor does not have an on-chip analog to digital converter (ADC), the data suffers from noise introduced in the signal line between the sensor and ADC chip. On-chip ADC circuitry is required to use the image sensor for high-resolution neural recording.

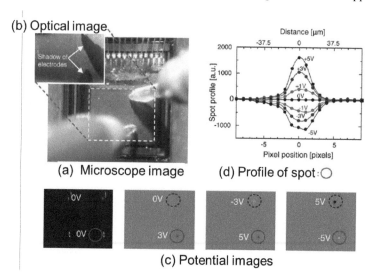

FIGURE 5.38: Experimental setup and results of potential imaging. (a) Micro-photograph of the measurement. Two gel-coated probes are placed on the sensor surface and potentials are applied. (b) On-chip optical image and (c) on-chip potential images. The profiles of the spots indicated by the solid circles in (c) are shown in (d). Adapted from [547] with permission.

5.4.2.2.2 Optical and electrochemical imaging Fluorescence is conventionally used to detect hybridized target DNA fragments on probe DNA spots, as described in the beginning of this section.   Electrochemical measurement is another promising detection scheme that can be an alternative or a supplementary method for micro-array technology [563–565]. Various electrochemical methods to detect hybridized bio-molecules have been proposed and some are being used in commercial equipment. Some groups have reported CMOS-based sensors for on-chip detection of bio-molecules using electrochemical detection techniques [549, 564, 566].

Design of sensor Figure 5.39 shows micro-photographs of a smart CMOS image sensor for optical and electrochemical dual imaging. The sensor was fabricated in 0.35-μm, 2-poly, 4-metal standard CMOS technology. It consists of a combined optical and electrochemical pixel array and control/readout circuitry for each function. The combined pixel array is a 128 × 128 light-sensing pixel array, partly replaced with electrochemical sensing pixels. The light-sensing pixel is a modified 3T-APS with a pixel size of 7.5 μm× 7.5 μm. The electrochemical-sensing pixel consists of an exposed electrode with an area of 30.5 μm× 30.5 μm using a transmission-gate switch for row selection. The size of the electrochemical-sensing pixel is 60 μm× 60 μm. Thus, 8 × 8 light-sensing pixels are replaced by one

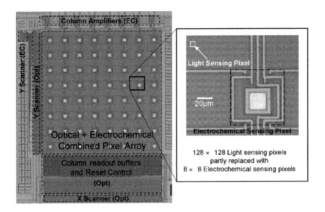

FIGURE 5.39: Micro-photograph of a fabricated smart CMOS sensor for optical and electrochemical dual imaging. Adapted from [549] with permission.

electrochemical-sensing pixel. The sensor has an 8×8 electrochemical pixel array embedded in the optical image sensor. Owing to the large mismatch in the operating speed between the optical image sensor and the electrochemical image sensor, the optical and electrochemical pixel arrays are designed to work independently of each other. Figure 5.40 shows the schematics of the sensor.

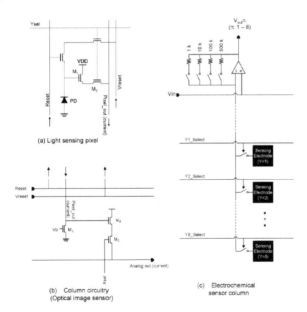

FIGURE 5.40: Pixel circuits for (a) optical sensing and (b) electrochemical sensing and (c) column circuits. Adapted from [549] with permission.

A voltage-controlled current measurement approach can be used for electrochemical measurements for on-chip bimolecular micro-array technology. Options include cyclic voltammetry (CV) [565] and differential pulse voltammetry [563], which have been reported to be feasible for detecting hybridized DNA. A current-sensing voltage follower is used for on-chip, multi-site electrochemical measurements. By inserting a resistance in the feedback path of the voltage follower (unity-gain buffer), the circuitry can perform voltage-controlled current measurements. This circuit configuration has been widely used in electrochemical potentiostats and patch-clamp amplifiers.

Experimental results   2D arrayed CV measurements have been performed and 8 × 8 CV curves were obtained using a single frame measurement. For on-chip measurements, gold was formed on the exposed aluminum electrodes of the electrochemical sensing pixels. Because of its chemical stability and affinity to sulfur bases, gold has been a standard electrode material for electrochemical molecular measurements. Au/Cr (300 nm/10 nm) layers were evaporated and patterned into the 30.5 μm × 30.5 μm electrochemical sensing electrodes. The sensor was then mounted on a ceramic package and connected with aluminum wires. The sensor with connecting wires was molded with an epoxy rubber layer. Only the combined pixel array was kept uncovered and exposed to the measurement solution.

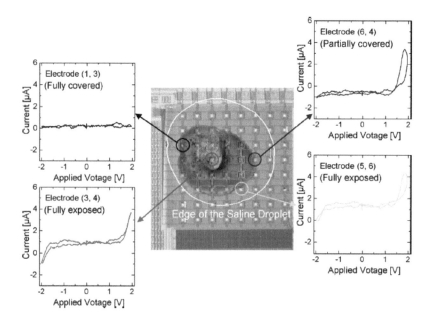

FIGURE 5.41: Experimental results of 2D arrayed CV measurements. Adapted from [549] with permission.

A two-electrode configuration was used for the arrayed CV measurements. An Ag/AgCl electrode was used as the counter electrode. The work electrode was an 8×8-array gold electrode. As a model subject for the 2D CV measurement, an agarose gel island with high resistivity in a saline solution was used. 8×8 CV curves were measured to take images of the electrochemical characteristics. The potential of the Ag/AgCl counter electrode was cyclically scanned between -3 and 5 V for each electrochemical row with a scan speed of 1 Hz. Figure 5.41 shows the results of the arrayed CV measurement. The observed CV profiles show different features depending on the condition of the individual measurement electrodes.

5.4.2.2.3 Optical and pH imaging

Design of sensor The pH imaging device describe in Sec. 5.3.2.2 is modified to detect optical image simultaneously. As the detection principle of PG image sensors described in Sec. 2.3.2 is similar to that of the pH imaging devices, devices with functions of pH and optical imaging have been developed [567–570]. Figure 5.42 shows the structure and operation principle of the device that can detect optical and pH images simultaneously. In this device, two FDs are placed on both sides; one is for pH detection and the other for optical detection as shown in the two cross-sectional views of the device in Figs. 5.42 (a) and (f). In the pH imaging mode (Figs. 5.42(b) – (e)), the charge is injected into the well, which is modulated by the density of proton or pH, while in the optical imaging mode (Figs. 5.42(f) – (k)), photo-generated charges are accumulated in the modulated well. It should be noted that in the pH imaging mode, charges are used to measure the well depth, while in the optical imaging mode, the amounts of charges are to be measured.

Experimental results Figure 5.43 shows the experimental results obtained by using the device [568]. In this figure, rice grain A, dipped in a buffer solution of pH 9.18 and rice grain B, dipped in a buffer solution of pH 4.01, are placed in the sensing area and measured in the pH and optical modes. The interval time between pH and optical modes is 0.2 ms. The pH images of the two grains can be distinguished, while the optical images are almost identical.

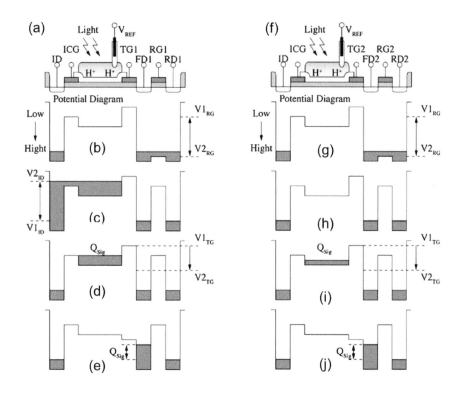

FIGURE 5.42: Structure and operating principle of pH and optical image sensor. Adapted from [568] with permission.

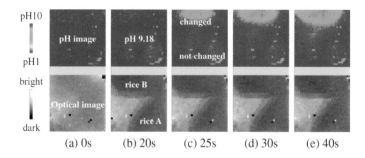

FIGURE 5.43: Experimental results of pH and optical image sensor. Adapted from [568] with permission.

5.4.3 Implantation type

5.4.3.1 *In vivo* brain imaging by using implantable smart CMOS image sensors

In this section, image sensors that are implanted in animals for bioscience/technology are described. The target organ for the implantation is the brain, especially the brain of experimental small animals such as mice and rats. To measure the biological functions in the brain optically, intrinsic optical signal and fluorescence are widely used. The direct imaging of blood vessels has also been carried out. Current technology for imaging the brain requires expensive equipment that has limitations in terms of image resolution and speed or imaging depth, which are essential for the study of the brain [571], especially of experimental small animals. As shown in Fig. 5.44, a miniaturized smart CMOS image sensor (denoted as "CMOS sensor" in the following discussion) is capable of real time *in vivo* imaging of the brain at arbitrary depths.

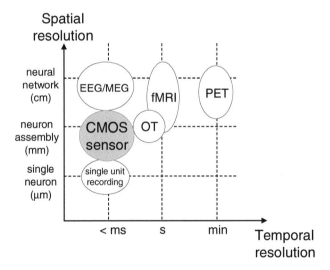

FIGURE 5.44: Current neural imaging technology. EEG: electroencephalography, MEG: magnetoencephalography, OT: optical topography, fMRI: functional magnetic resonance imaging, and PET: positron emission tomography.

5.4.3.1.1 Image sensor implementation The implantable imaging devices should be so small as to be implanted in the brain of mice, cause little damage to the brain, and work in long term for behavior experiments. To meet these requirements, implantable miniaturized image sensing devices have been developed. The device is integrated with a dedicated CMOS image sensor and LEDs on a flexible substrate. The CMOS sensor is based on a 3T-APS with a modification for reducing the number of inputs and outputs (IOs) as shown in Fig. 5.45.

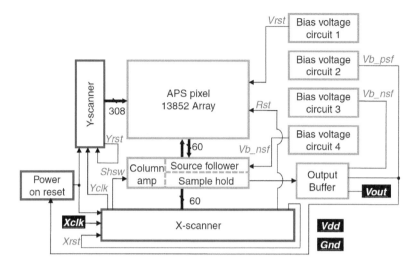

FIGURE 5.45: Chip block diagram of a smart CMOS image sensor for *in vivo* imaging. There are only four IOs. Illustrated after [572].

The number of IO pads are only four; V_{dd}, GND, Clock signal and V_{out}, and the pads are placed on one side of the chip so that it can be easily implanted in the brain [572]. Two types of implantable image sensors have been developed for brain surface and deep brain insertion as shown in Fig. 5.46.

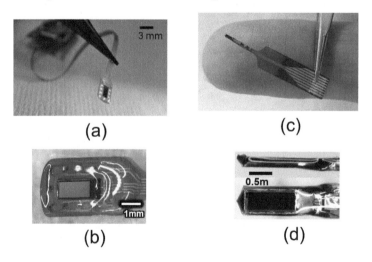

FIGURE 5.46: Photographs of implantable miniaturized imaging devices. (a) Brain surface type, and (b) its close-up photo. (c) Needle-type, and (d) its close-up photo. Adapted from [573] with permission.

The specifications of these two types are listed in Table 5.2. It should be noted that the sensors have only an imaging function, though it is possible to integrate the sensors with other functions such as extracellular potential measurement and optical stimulation, which are introduced later in this section.

TABLE 5.2
Specifications of a smart CMOS image sensor for *in vivo* brain imaging

Type	Planar shape	Needle shape
Implant place	Brain surface	Deep brain
Number of pixels	120×268	40×120
Chip size	1×3.5 mm^2	0.45×1.5 mm^2
Image area	0.9×2.01 mm^2	0.3×0.9 mm^2
Pixel size	7.5×7.5 μm^2	
Fill factor	29%	
Technology	0.35-μm CMOS 2P3M	
Pixel type	Modified 3T-APS	
Photodiode	N-well/P-substrate junction	
Number of IOs	4 (V_{dd}, GND, Clock, V_{out})	

Glucose sensing by implantable micro-imaging device   Another application of implantable fluorescence imaging devices other than brain imaging is glucose sensing. These devices use a special gel which emits fluorescence under glucose. By combining the gel with a micro-image sensor similar to the image sensor previously described, glucose sensing has been achieved [574–576]. In the next section, implantable CMOS image sensors for *in vivo* imaging of intrinsic optical signal and fluorescence in mouse are presented.

5.4.3.1.2 *in vivo* intrinsic optical signal imaging of rodent brain   In this section, implantable imaging devices that observe *in vivo* intrinsic optical signal imaging are introduced. To measure the intrinsic optical imaging in a rodent brain, miniaturized CMOS image sensors are implanted on the surface of the rodent brain directly. The sensor specifications are described in the previous section. The implantable device consists of a CMOS image sensor and LEDs for observing blood as an optical intrinsic signal source on a flexible substrate. Green-colored LEDs are used for observing the blood flow. The light from a green-colored LED does not deeply penetrate the brain tissue and can observe only blood vessels near the brain surface. In Fig. H.2 of Appendix H, the absorption spectrum in the tissue is shown. The "*in vivo* window" is the wavelength region where the light around the wavelength can penetrate the tissue deeply owing to low absorption of both water and hemoglobin.

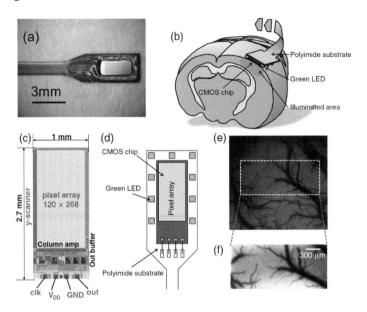

FIGURE 5.47: *In vivo* intrinsic optical signal imaging [577]. (a) Photograph of the whole device, (b) illustration of the experimental setup, (c) CMOS image sensor, (d) device structure (one CMOS image sensor and nine green LEDs (λ=535 nm) on a flexible polyimide substrate), (e) photograph of the brain surface with optical microscope, and (f) image taken by the implanted device. Adapted from [577] with permission.

The brain activity of neurons depends on a continuous supply of oxygen and glucose through the blood. Thus, the absorption of the green light in the local region increases with an increase in blood volume. This is called hemodynamic signals indicating blood-flow changes in the brain.

Figure 5.47 shows the concept and the blood vessel image taken by the device implanted on the surface of a rat [577]. The green LEDs (λ=535 nm) are placed around the sensor as shown in Fig. 5.47(d) so as to achieve uniform illumination on the brain surface. The blood vessels are clearly observed by the device.

In the experiment, the responses of the whisker region in the somato-sensory cortex is measured by stimulating the right whiskers using the implanted device mentioned above. Figure 5.48 shows the experimental results. They demonstrate that the implantable device can clearly measure the response evoked by the physiological stimulation in the rat's brain.

The device can detect the brain activity through the blood flow in blood vessels. The other method is to use intrinsic optical signal, especially hemoglobin conformation change. The largest change of hemoglobin conformation occurs around 600-700 nm as shown in Fig. H.1 in Appendix H. Thus, if the device has red-colored LEDs as well as green-colored LEDs, it can detect hemoglobin conformation change as well as blood flow change [578].

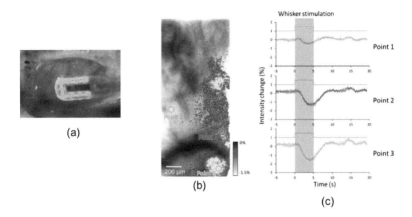

FIGURE 5.48: Result of the intrinsic optical signal imaging with whisker stimulation. (a) Implanted device on the brain surface. (b) Image of the intrinsic signal responses of the right-whisker region of the left primary somato-sensory cortex. (c) Results of the whisker reactions. When the right whisker is stimulated, the reactions of the intrinsic optical signal are obtained. At point 3, strong reactions were obtained. Adapted from [579] with permission.

5.4.3.1.3 *in vivo* **rodent brain imaging of fluorescence** Fluorescence contains important information in bioscience/technology measurements. In fluorescence measurements, excitation light is typically used, and thus fluorescence can be distinguished as a signal light from the background signal owing to the excitation light. The intensity of fluorescence light is much weaker than that of the excitation light, with the proportion typically being about 1%. To suppress the background light, on-chip color filters for the rejection of excitation light [573, 580] and potential profile control [285, 286] have been proposed and demonstrated. Potential profile control is explained in Sec. 3.3.5.3.

An important application here is the imaging of the rodent's brain activity. Packaging is a critical issue for applications where the sensor is inserted deeply into the mouse brain. Figure 5.49 illustrates a packaged device, which is integrated with a smart CMOS image sensor and one or two LEDs for the excitation of fluorescence on a flexible polyimide substrate. A dedicated fabrication process was developed, which makes it possible to realize an extremely compact device measuring 150 μm in thickness. Another experiment shows that minimal injury was inflicted on the brain using this method as the brain continued to function and respond normally with the device inside it.

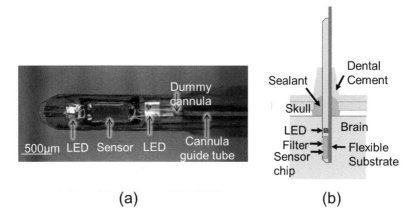

(a)   (b)

FIGURE 5.49: Implantable needle-type imaging device: (a) device photograph, and (b) cross-section of implantation of the device into the rodent brain. Adapted from [573] with permission.

Figure 5.50 shows experimental results by using the implantable imaging device [573]. The device was implanted in the deep brain region, CA1 in hippocampus of a mouse. The mouse was induced with an artificial epilepsy and showed seizure. The fluorescence intensity corresponds to the neural activities in CA1 and coincides with the temporal change in the level of seizure such as nodding as shown in Fig. 5.50(c) and clonic convulsion as shown in Fig. 5.50(d).

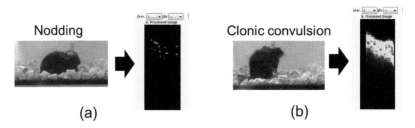

(a)   (b)

FIGURE 5.50: Experimental results from *in vivo* deep brain imaging device implanted into the hippocampal CA1 region of a mouse and measurement result of brain activities by artificially inducing epilepsy. The behavior of nodding (a) and seizure (b) is shown with corresponding brain images taken by the implanted device. Adapted from [573] with permission.

5.4.3.1.4 Self-reset imaging The issue of intrinsic optical signal and fluorescence detection in this implantable application is discussed here. The background light in fluorescence and intrinsic signal is quite strong, and its change is very small. For example, the change of fluorescence is typically about 1%. As a result, sometimes the photodiode is saturated due to the fluorescence or intrinsic light. To solve this problem, an implantable image sensor with self-reset mode, the operation principle of which is described in Sec. 3.2.2.2.1 of Chapter 3, has been developed [228, 229]. In Fig. 5.51, it is observed that the response of electrical stimulation through the forelimb and hindlimb can be obtained by measuring the response of the brain blood. According to the stimulation, it is observed that the response originates in the forelimb region in the brain and not in the hindlimb region. By using this self-reset type CMOS image sensor, the stimulation response is clearly obtained.

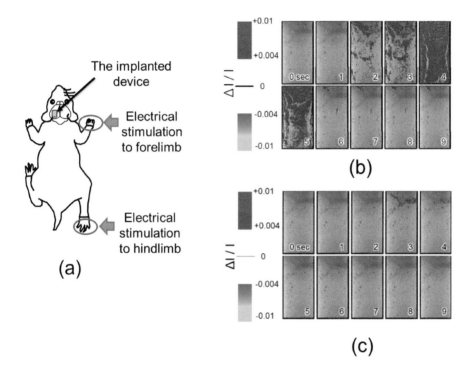

FIGURE 5.51: Experimental results of intrinsic optical imaging in vivo. (a) the experimental setup of the sensor and electrical stimulation points, (b) time lapse images for electrical stimulation to forelimb, and (c) to hindlimb. Adapted from [229] with permission.

5.4.3.2   Implantable image sensor with multi-modal function

5.4.3.2.1   Implantable image sensor with electrical recording   In this section, recording electrodes are integrated on a chip. Figure 5.52 shows an *in vivo* smart CMOS image sensor with an LED array as excitation source and recording electrodes [572]. When metal electrodes are placed on an imaging area, input light cannot hit on the PD covered with metal electrodes. To alleviate this issue, a mesh-type metal electrode is introduced as shown in Fig. 5.52. In this figure, the close-up photo of one pixel is shown, where the aluminum electrode metal is open on the top of the PD. The captured image in Fig. 5.52 demonstrates that the electrodes have very little effect on the image.

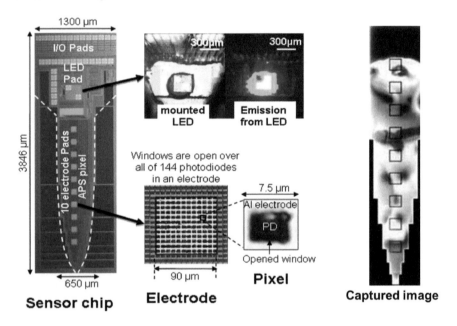

FIGURE 5.52: Advanced smart CMOS image sensor for *in vivo* imaging integrated with electrodes. Adapted from [572] with permission.

5.4.3.2.2   Implantable image sensor with optical stimulation   In the previous section, the device can be used as electrical recording, and it can also be used as an electrical stimulation to neural cells if the electrode materials are changed to the ones suitable for electrical stimulation in electrolytes such as Pt instead of Al. Recently, another stimulation method has emerged known as optogenetics [581]. Optogenetics is a genetic method by expressing opsins in cells. The cell expressed opsin is modified to have photosensitive property such as being evoked by light illumination. Widely used opsin is Channel Rhodopsin 2 (ChR2). By combining an image sensor with an optical stimulation array device, we can observe biological

activities when neural cells are activated by light pattern [582–584]. In addition, an optical bi-directional communication with cells can be established by achieving feedback loop with optical stimulation and fluorescence measurement [585].

Figure 5.53(a) shows that a blue-LED array (λ=480 nm) is stacked on a CMOS image sensor integrated with control circuits through metal bumps and green LEDs to observe hemodynamic signals [584]. The image sensor is integrated with electrical control circuits for the LED array. The block diagram of the chip is shown in Fig. 5.53(b). As shown in Fig. 5.53(c), the two chips of the CMOS image sensor and the LED array are stacked by the flip-chip bonding method through Au bumps.

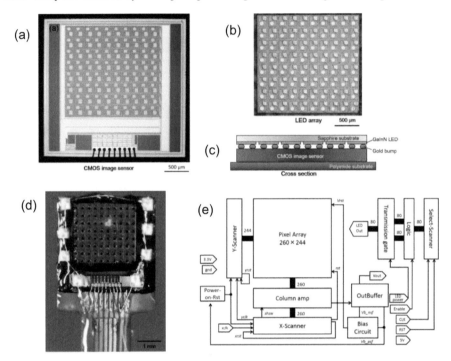

FIGURE 5.53: Implantable optogenetic device: (a) chip photo of CMOS image sensor with LED array control circuits, (b) photo of LED array, (c) schematic cross-section of the device, (d) fabricated device, and (e) block diagram. Adapted from [584] with permission.

By using this device, the brain surface of a transgenic mouse expressed ChR2 is optically stimulated. The optical stimulation with the blue-LED array (λ=480 nm), which is confirmed by electrophysiological measurement of evoked cells, activates hemodynamic signals indicating blood flow change in the brain. The blood vessels can be visualized by the imaging function of the device under the illumination of green LEDs as shown in Fig. 5.54. It should be noted that the blue-LED array is transparent except for the electrodes so that almost the whole image can be taken by the device as shown in Fig. 5.54(d).

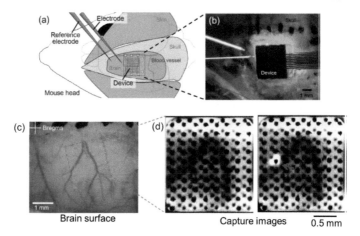

FIGURE 5.54: Experimental setup and results. (a) Device placement on the brain surface of the transgenic mouse; the electrodes for electrophysiology are also shown. (b) Micro-photograph of the mouse head after the setup of the device. (c) Micro-photograph of the brain surface; a square frame indicates the region from which the implantable device captured the images. (d) Image captured using the device; the black dots show the electrodes of the LED array; the arrow in the right photo indicates the turn-on LED. Adapted from [584] with permission.

5.5 Medical applications

In this section, two medical applications for smart CMOS image sensors are presented, namely: capsule endoscopes and retina prosthesis. Smart CMOS image sensors are suitable for medical applications for the following reasons. Firstly, they can be integrated with signal processing, RF, and other electronics, that is, a system-on-chip (SoC) is possible. This can be applied to a capsule endoscope, which requires a system with small volume and low power consumption. Second, smart functions are effective for medical use. Retinal prosthesis is one such example requiring an electronic stimulating function on a chip. In the near future, the medical field will be one of the most important application areas for smart CMOS image sensors.

5.5.1 Capsule endoscope

An endoscope is a medical instrument for observing and diagnosing organs such as the stomach and intestines by being inserted into the body. It is employed with a CCD/CMOS camera with a light-guided glass fiber to illuminate the area being observed. An endoscope or push-type endoscope is a highly integrated instrument with a camera, light guide, small forceps to pick up a tissue, a tube for injecting water to clean the tissues, and an air tube for enlarging the affected region. A capsule endoscope is a type of endoscopes developed in 2000 by Given Imaging in Israel [586]. Olympus has also commercialized a capsule endoscope. Figure 5.55 shows a photograph of Olympus's capsule endoscope.

FIGURE 5.55: Capsule endoscope. A capsule with a length of 26 mm and a diameter of 11 mm is used with a dome, optics, white LEDs for illumination, CMOS camera, battery, RF electronics, and antenna. The logo in the photo is not printed in the commercial products. By courtesy of Olympus.

A capsule endoscope uses a CMOS image sensor, imaging optics, white LEDs for illumination, RF circuits, antenna, battery, and other elements. A user swallows a capsule endoscope and it automatically moves along the digestive organs. Compared with a conventional endoscope, a capsule endoscope causes less pain to the patient. It should be noted that a capsule endoscope is limited to usage in the small intestine, and is not used for the stomach or large intestine (colon). Recently, capsule endoscopes for observing the esophagus [587] and the colon [588] have been developed. They have two cameras to image forward and rear areas.

Smart CMOS image sensor for capsule endoscope A capsule endoscope is a type of implanted device, and hence the critical issues are size and power consumption. A smart CMOS image sensor is thus suitable for this purpose. When applying a CMOS image sensor, color realization must be considered. As discussed in Sec. 2.10, a CMOS image sensor uses a rolling shutter mechanism. Medical use generally requires color reproducibility and hence, three image sensors or three light sources are preferred, as discussed in Sec. 2.8. In fact, some conventional endoscopes use the three-light sources method. For installing a camera system in a small volume, the three-light sources method is particularly suitable for a capsule endoscope. However, owing to the rolling shutter mechanism, the three-light sources method cannot be applied to CMOS image sensors. In a rolling shutter, the shutter timing is different in every row, and hence the three-light sources method, in which each light emits at different timing cannot be applied. Present commercially available capsule endoscopes use on-chip color filters in CMOS image sensors. To apply the three-light sources method in a CMOS image sensor, a global shutter is required. Another method has been proposed that calculates the color reproducibility. As the RGB mixing ratio is known a priori in a rolling shutter when using the three-light source method, RGB can be separated by calculating outside the chip [589].

Because a capsule endoscope operates through a battery, the power consumption of the total electronics should be small. Furthermore, the total volume should be small. Thus, SoC for the imaging system including the RF electronics is effective. For this purpose, SoC with a CMOS image sensor and RF electronics has been reported [590]. As shown in Fig. 5.56, the fabricated chip has only one I/O pad of a digital output besides a power supply (Vdd and GND) integrated with BPSK (Binary Phase Shift Keying) modulation electronics. The chip consumes 2.6 mW under a condition of 2 fps with QVGA format. A SoC for a capsule endoscope has been reported, though the image sensor is not integrated. This system has the capability of wirelessly transmitting data of 320 × 288 pixels in 2 Mbps with a power consumption of 6.2 mW [591].

Another desirable function of a capsule endoscope is the on-chip image compression. There are several reports of on-chip compression [592–595] and in the near future, it is expected that this function will be employed by a capsule endoscope. These SoCs will be used in capsule endoscopes as well as combining with technologies such as micro-electro-mechanical systems (MEMS), micro-total analysis system (μTAS), and lab-on-chip (LOB) to monitor other physical values such as potential, pH, and temperature [596, 597]. Such multi-modal sensing is suitable for the smart CMOS image sensor described in the previous section, Sec. 5.4.2.2.

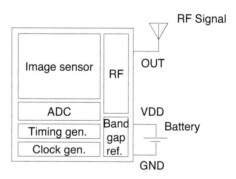

FIGURE 5.56: SoC including a CMOS image sensor for capsule endoscopes [590]. ADC: analog-to-digital converter. Timing gen.: timing pulse generator. Clock gen.: internal clock generator. A cyclic ADC is used in this chip. Illustrated after [590].

5.5.2 Retinal prosthesis

In an early work in this field, MOS image sensors were applied to help the blind. The Optacon, or optical-to-tactile converter, is probably the first use of a solid-state image sensor for the blind [598]. The Optacon integrated scanning and readout circuits and was compact in size [120, 599]. Retinal prosthesis is like an implantable version of the Optacon. In the Optacon, the blind perceive an object through tactile sense, while in retinal prosthesis the blind perceive an object through an electrical stimulation of vision-related cells by an implanted device.

A number of studies have been carried out considering different implantation sites such as cortical, epi-retina, sub-retina and suprachoroid places. Implantation in the retinal space or ocular implantation prevents infection and can be applied to patients suffering from retinitis pigmentosa (RP) and age-related macular degeneration (AMD) where retinal cells other than the photo-receptors still function. It is noted that both RP and AMD are diseases with no effective remedies yet. The structure of the retina is shown in Fig. C.1 of Appendix C.

While in the epi-retinal approach, ganglion cells (GCs) are stimulated, in the sub-retinal approach, the stimulation is merely a replacement of the photo-receptors, and thus in an implementation of this approach, it is likely that the bipolar cells as well as the GCs will be stimulated. Consequently, the sub-retinal approach has the following advantages over the epi-retinal approach: there is little retinotopy, that is, the stimulation points correspond well to the visual sense, and it is possible to naturally utilize the optomechanical functions such as the movement of the eyeball and opening and closing of the iris. Figure 5.57 illustrates the three methods of epi- and sub-retinal stimulations and suprachoroid transretinal stimulation (STS).

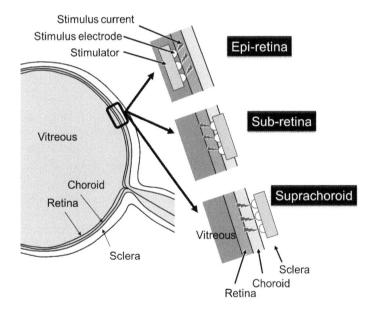

FIGURE 5.57: Three methods of retinal prosthesis for ocular implantation.

5.5.2.1 Intraocular implantation of CMOS devices

Figure 5.58 shows the classification of imaging systems in retinal prosthesis. Although intraocular implantation approaches have advantages over other approaches, it should be noted that there are many technical challenges to overcome

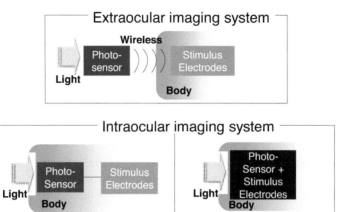

FIGURE 5.58: Imaging system in retinal prosthesis.

when applying CMOS-based simulator devices to retinal prosthesis. Compared with the epi-retinal approaches, the sub-retinal approach has more difficulties in using the CMOS chips because it is completely implanted into the tissues and must work for both the image sensors and the electric simulators. We need to consider the following points for intraocular implantation of CMOS devices.

Issues to consider for intraocular implantation of CMOS devices A CMOS-based interface must be bio-compatible. The standard CMOS structure is unsuitable for a biological environment; silicon nitride is conventionally used as a protective top layer in standard CMOS process; however, it will be damaged in a biological environment when implanted for a long time.

The stimulus electrodes must be compatible with the standard CMOS structure. Wire-bonding pads, which are made of aluminum, are typically used as input–output interfaces in standard CMOS technology, but are completely inadequate as stimulus electrodes for retinal cells, because aluminum dissolves in a biological environment. Platinum is a candidate for stimulus electrode materials. These issues are discussed in detail in Ref. [223].

5.5.2.2   Epi-retinal approach using an image sensor device

It should be noted that the sub-retinal approach is natural when using imaging with stimulation because imaging can be done on the same plane of stimulation. Some epi-retinal approaches, however, can use implanted imaging device with stimulation. As mentioned in Sec. 3.3.1.1, silicon-on-sapphire (SOS) is transparent and can be used in an epi-retinal approach by a BSI image sensor. For the back-illuminating configuration, the imaging area and stimulation can be placed on the same plane. A PFM photo-sensor using SOS CMOS technology has been demonstrated for the epi-retinal approach [214].

Another epi-retinal approach using an image sensor is to use 3D integration technology [224, 600, 601]. Figure 5.59 shows the concept of retinal prosthesis using 3D integration technology. By introducing the present advanced stacked technology, more efficient retinal prosthesis devices could be developed.

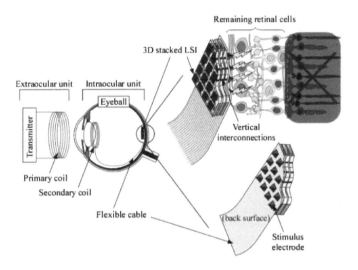

FIGURE 5.59: Epi-retinal approach using 3D integration technology [224]. Courtesy of Prof. M. Koyanagi.

5.5.2.3 Sub-retinal approach

In sub-retinal implantation, a photo-sensor is required to integrate the stimulus electrode. Figure 5.60 shows the classification of retinal stimulator in which electrical stimulators and photo-detectors are integrated. A simple photodiode array without any bias voltage, that is, a solar cell mode, has been used as a photo-sensor owing to its simple configuration (no power supply is needed) [602–604]. The photocurrent is directly used as the stimulus current into the retinal cells. Because the direct photocurrent under normal illumination condition is not enough to evoke retinal cells, this type of stimulator cannot be used as retinal stimulator as it is.

Photocurrent under normal illumination condition In the discussion of Sec. 2.4 of Chapter 2, we observe that a photocurrent of about 10 pA is produced from a PD with the area size of 100 μm^2 and a photo sensitivity of 0.6 A/W under 1,000 lux lighting condition, which indicates a little larger value than that in a normal room. If the area size of each PD in a retinal stimulator based on micro-solar cell is 100 × 100 μm^2, the photocurrent is about 1 nA. This value is not enough to evoke retinal cells directly.

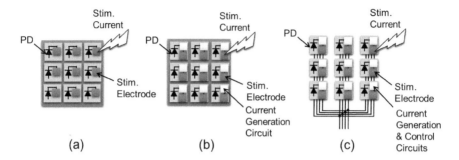

FIGURE 5.60: Three configurations of sub-retinal implantation. (a) Micro-solar array. PD is directly connected to the stimulus electrode. (b) Retinal stimulator based on CMOS image sensor. Each pixel has a stimulus electrode. (c) Micro-chip based retinal stimulator; each micro-chip is integrated with a PD and a stimulus electrode as well as circuits for current generation and control; it has its own ID number so as to be assigned by external control signal.

Micro-solar cell array   Consequently, retinal stimulators based on solar cell array should be modified by enhancing the photocurrent enough to evoke retinal cells, that is, over μA. Figure 5.61 shows a sub-retinal stimulator with a tandem PD structure [605, 606]. Three PDs are connected serially to produce the voltage to stimulate the retinal cells adequately. In addition, the image taken by an external camera is converted into NIR image the intensity of which is amplified electrically. This amplified NIR image illuminates the PDs so that the photocurrent increases and reaches μA order.

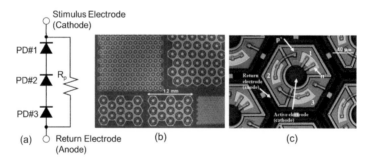

FIGURE 5.61: Retinal stimulator based on micro-solar array. (a) Equivalent circuits. Three PDs are connected serially to produce voltage to stimulate the retinal cells adequately. (b) SEM photograph of pixel arrays. (c) Micro-photograph of a pixel. Adapted from [606] with permission.

Figure 5.62 is another type of retinal sub-retinal stimulator based on the solar cell mode as shown in Fig. 5.62. This configuration is similar to a CMOS image sensor.

PDs as solar cells are placed in the peripheral area of the chip. This chip works based on Divisional Power Supply Scheme (DPSS) [607]. In DPSS, all the pixels are divided into N-group and when one group operates as image sensing, other groups work as solar cells so as to provide photo-generated power to the group. Natural image is converted into an intensified NIR pattern to supply enough power to drive the chip.

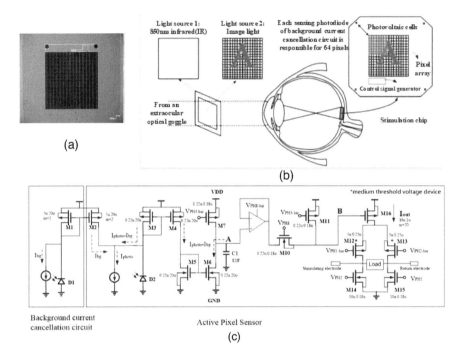

FIGURE 5.62: Sub-retinal stimulator chip integrated with micro-solar cell array based on DPSS for efficient power supply. Adapted from [607] with permission.

CMOS image sensor based stimulator The third method for sub-retinal stimulator is to use a CMOS image sensor. In each pixel, a stimulus electrode is placed. The amount of stimulus current is enough to evoke retinal cells because of in-pixel amplifier circuits. Figure 5.63 shows the pixel circuits of the stimulator [608], which has been developed by Zrenner's group in Germany. The photo-sensor is composed of logarithmic photo-sensor, which is described in Sec. 3.2.1.2 of Chapter 3, so that the sensor can achieve a wide dynamic range of 7–decade over the input light. It should be noted that the image quality of the logarithmic photo-sensor is not very good conventionally, however, it is sufficient in this application. The photo-current is amplified by the differential amplifier where another input current produced by global illumination is differentiated. The differential current is transferred by mirror circuits and the final current direction is determined based on which power line of V_H and V_L is positive. It should be noted that the power of this chip is supplied from external body and the chip is directly implanted in the retina so that the electrical cable from the implanted coil to the stimulator chip must be driven in the AC mode instead of conventional DC mode to avoid the risk of electrolysis when the cable is broken [608, 609].

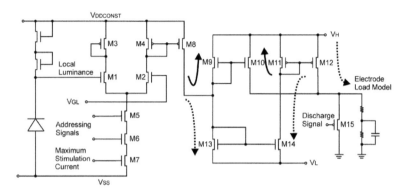

FIGURE 5.63: Pixel circuits of sub retinal stimulator with CMOS image sensor based stimulator. Adapted from [608] with permission.

Multiple micro-chip based stimulator   The previous chips are mainly CMOS chips and, therefore, they may block the flow of nutrition from epithelium to retinal cells. In addition, the chip is too hard to bend along the eyeball. To overcome these issues, multiple micro-chip architecture has been proposed [223, 610] as shown in Fig. 5.60(c). Figure 5.64 shows the chip based on multiple micro-chip architecture. The stimulation timing is determined with the the pulse width of CONT in Fig. 5.64(c).

The fabricated chip is implanted on the suprachoroidal place as shown in Fig. 5.65(a). NIR light is used to input the implanted chip and not to evoke the retinal cells. The experimental results are shown in Fig. 5.65(b). It is clearly demonstrated that only under light illumination condition, the retinal cells are evoked.

Based on the architecture above, flexible stimulator is fabricated as shown in Fig. 5.66. Figure 5.66(a) shows a mother chip that contains unit chips before dividing. In the mother chip, three unit chips are placed along 16 radial directions and one is placed in the center, and thus the total number of chips is 49 as shown in Fig. 5.66(a). The unit chips are placed on a flexible substrate as shown in Fig. 5.66(c), and the experimental results are shown in Fig. 5.66 (f), where a patterned light "N" inputs the device and the output current pattern from the device corresponds to the input light pattern. Thus, this device can sense the image and output the corresponding stimulus current pattern.

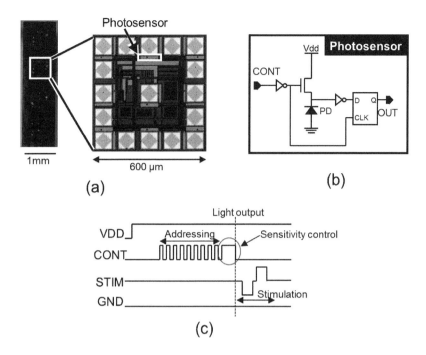

FIGURE 5.64: Multiple micro-chip for retinal stimulation: (a) chip photo, (b) chip circuit, and (c) timing chart. Adapted from [611] with permission.

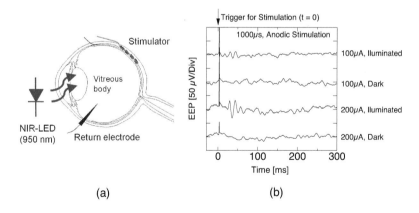

FIGURE 5.65: In vivo experiments to validate the function of multiple micro-chip architecture for retinal stimulation: (a) Implantation place, (b) Electric evoked potential (EEP) signals by stimulation with the implanted chip. Adapted from [611] with permission.

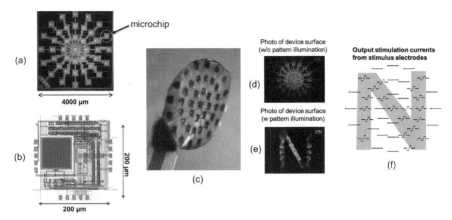

FIGURE 5.66: Retinal stimulator based on multiple micro-chip architecture: (a) a mother chip that contains 49 unit chips before dividing, (b) the layout of the unit chip, (c) the fabricated retinal stimulator, (d) the photograph of device surface without pattern illumination, (e) the same in (d) without pattern "N" illumination, and (f) experimental results where output stimulation currents from stimulus electrodes. Adapted from [612] with permission.

PFM photo-sensor based stimulator To generate sufficient stimulus current using a photo-sensor in a daylight environment, a pulse frequency modulation (PFM) photo-sensor has been proposed for the sub-retinal approach [212, 223]. Furthermore, PFM-based retinal prosthesis devices as well as a simulator for STS have been developed [202, 205, 213–222, 613–617].

Several groups have also developed a PFM photo-sensor or pulse-based photo-sensor for use in sub-retinal implantation [224–227, 600, 601, 618, 619].

The PFM appears to be suitable as a retinal prosthesis device in sub-retinal implantation for the following reasons. Firstly, the PFM produces an output of pulse streams, which would be suitable for stimulating the cells. In general, pulse stimulation is effective for evoking cell potentials. In addition, such a pulse form is compatible with logic circuits, which enables highly versatile functions. Secondly, PFM can operate at a very low voltage without decreasing the SNR. This is suitable for an implantable device. Finally, its photosensitivity is sufficiently high for detection in normal lighting conditions and its dynamic range is relatively large. These characteristics are very advantageous for the replacement of photo-receptors. Although the PFM photo-sensor is essentially suitable for application to a retinal prosthesis device, some modifications are required and these will be described herein.

Modification of PFM photo-sensor for retinal cell stimulation To apply a PFM photo-sensor to retinal cell stimulation, the PFM photo-sensor must be modified. The reasons for the modification are as follows.

- Current pulse: The output from a PFM photo-sensor has the form of a voltage pulse waveform, whereas a current output is preferable for injecting charges into retinal cells, even if the contact resistances between the electrodes and the cells are changed.
- Biphasic pulse: Biphasic output, that is, positive and negative pulses, is preferable for charge balance in the electrical stimulation of retinal cells. For clinical use, charge balance is a critical issue because residue charges accumulate in living tissues, which may cause harmful effects to retinal cells.
- Frequency limit: An output frequency limitation is needed because an excessively high frequency may cause damage to the retinal cells. The output pulse frequency of the original PFM device shown in Fig. 3.9 is generally too high (approximately 1 MHz) to stimulate the retinal cells. The frequency limitation, however, causes a reduction in the range of the input light intensity. This problem is alleviated by introducing a variable sensitivity wherein the output frequency is divided into 2^{-n} portions with a frequency divider. This idea is inspired by the light-adaptation mechanism in animal retina, as illustrated in Fig. C.2 of Appendix C. Note that the digital output of the PFM is suitable for the introduction of the logic function of the frequency divider.

Design of sensor Based on the above modifications, a pixel circuitry has been designed and fabricated using standard 0.6-μm CMOS technology [213]. Figure 5.67 shows a block diagram of the pixel. Frequency limitation is achieved by a low-pass filter using switched capacitors. A biphasic current pulse is implemented by switching the current source and sink alternatively.

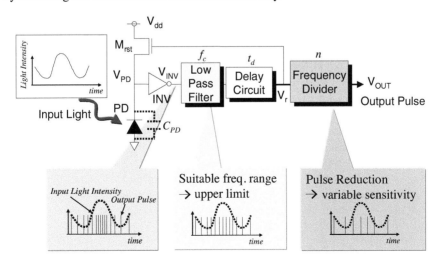

FIGURE 5.67: Block diagram of PFM photo-sensor pixel circuit modified for retinal cell stimulation. Illustrated after [620].

Figure 5.68 shows experimental results of variable photosensitivity using the chip. The original output curve has a dynamic range of over 6-log (6th-order range of input light intensity), but is reduced to around 2-log to be limited at 250 Hz when the low-pass filter is turned on. By introducing variable sensitivity, the total coverage of input light intensity becomes 5-log between $n = 0$ and $n = 7$, where n is the number of divisions.

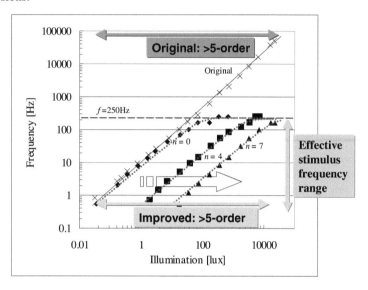

FIGURE 5.68: Experimental results of variable photosensitivity of the PFM photo-sensor. Illustrated after [213].

Application of PFM photo-sensor to the stimulation of retinal cells   In this section, the PFM-based stimulator described in the previous section is demonstrated to be effective in stimulating retinal cells. To apply the Si-LSI chip to electro-physiological experiments, we must protect the chip against the biological environment, and make an effective stimulus electrode that is compatible with the standard CMOS structure. To meet these requirements, a Pt/Au stacked bump electrode has been developed [614–617]. However, owing to the limited scope of this book, this electrode will not be described here.

To verify the operation of the PFM photo-sensor chip, *in vitro* experiments using detached bullfrog retinas were performed. In this experiment, the chip acts as a stimulator that is controlled by input light intensity, as is the case in photo-receptor cells. A current source and pulse shape circuits are integrated onto the chip. The Pt/Au stacked bump electrode and chip molding processes were performed as described in [614–617].

A piece of bullfrog retina was placed, with the retinal ganglion cell (RGC) side face up, on the surface of the packaged chip. Figure 5.69 shows the experimental

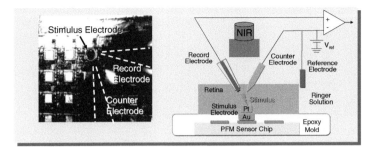

FIGURE 5.69: Experimental setup of *in vitro* stimulation using the PFM photo-sensor. Illustrated after [219].

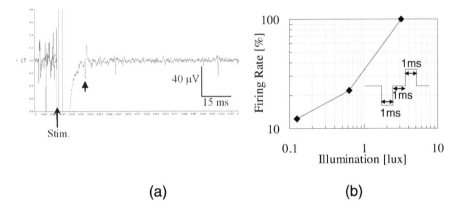

(a)   (b)

FIGURE 5.70: Experimental results of *in vitro* stimulation using the PFM photo-sensor: (a) example waveform obtained, and (b) firing rate as a function of input light intensity. Adapted from [219] with permission.

setup. Electrical stimulation was performed using the chip at a selected single pixel. A tungsten counter electrode with a tip diameter of 5 μm was placed on the retina, producing a trans-retinal current between the counter electrode and the chip electrode. A cathodic-first biphasic current pulse was used as the stimulant. The pulse parameter is described in the inset of Fig. 5.69. Details of the experimental setup are given in Ref. [219]. Note that the NIR light (near-infrared) does not excite the retinal cells, but it does excite the PFM photo-sensor cells.

Figure 5.70 also demonstrates the experimental results of evoking the retinal cells with the PFM photo-sensor, which is illuminated by input the NIR light. The firing rate increases in proportion to the input NIR light intensity. This demonstrates that the PFM photo-sensor activates the retinal cells through the input of NIR light, and suggests that it can be applied to human retinal prosthesis.

Appendices

A

Tables of constants

TABLE A.1

Physical constants at 300 K [79]

Quantity	Symbol	Value	Unit
Avogadro constant	N_{AVO}	6.02204×10^{23}	mol^{-1}
Boltzmann constant	k_B	1.380658×10^{-23}	J/K
Electron charge	e	$1.60217733 \times 10^{-19}$	C
Electron mass	m_e	$9.1093897 \times 10^{-31}$	kg
Electron volt	eV	$1 \text{ eV} = 1.60217733 \times 10^{-19}$ J	
Permeability in vacuum	μ_o	1.25663×10^{-6}	H/m
Permittivity in vacuum	ε_o	$8.854187817 \times 10^{-12}$	F/m
Planck constant	h	$6.6260755 \times 10^{-34}$	J·s
Speed of light in vacuum	c	2.99792458×10^{8}	m/s
Thermal voltage at 300 K	$k_B T$	26	meV
Thermal noise in 1 fF capacitor	$\sqrt{k_B T/C}$	5	μV
Wavelength of 1-eV quantum	λ	1.23977	μm

TABLE A.2

Properties of some materials at 300 K [79]

Property	Unit	Si	Ge	SiO_2	Si_3N_4
Bandgap	eV	1.1242	0.664	9	5
Dielectric constant		11.9	16	3.9	7.5
Refractive index		3.44	3.97	1.46	2.05
Intrinsic carrier conc.	cm^{-3}	1.45×10^{10}	2.4×10^{13}		
Electron mobility	cm^2/Vs	1430	3600	-	-
Hole mobility	cm^2/Vs	470	1800	-	-

B

Illuminance

Figure B.1 shows typical levels of illuminance for a variety of lighting conditions. The absolute threshold of human vision is about 10^{-6} lux [301].

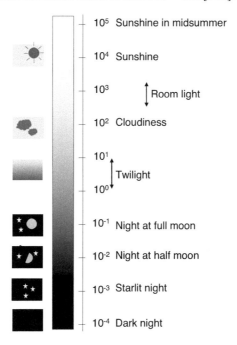

FIGURE B.1: Typical levels of illuminance for a variety of lighting conditions [6, 158].

Radiometric and photometric relation   Radiometric and photometric quantities are summarized in Table B.1 [158].

The response of a photopic eye $V(\lambda)$ is shown in Fig. B.2.

The conversion factor K from a photopic quantity to a physical quantity is expressed as

$$K = 683 \frac{\int R(\lambda)V(\lambda)d\lambda}{\int R(\lambda)d\lambda}. \tag{B.1}$$

Typical conversion factors are summarized in Table B.2 [158].

TABLE B.1

Radiometric quantities vs. photometric quantities K(lm/W) [158]

Radiometric quantity	Radiometric unit	Photometric quantity	Photometric unit
Radiant intensity	W/sr*	Luminous intensity	candela (cd)
Radiant flux	W=J/S	Luminous flux	lumen (lm) =cd·sr
Irradiance	W/m²	Illuminance	lm/m² = lux
Radiance	W/m²/sr	Luminance	cd/m²

*sr:steradian

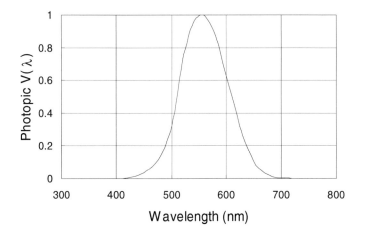

FIGURE B.2: Photopic eye response.

TABLE B.2

Typical conversion factors, K(lm/W) [158]

Light source	Conversion factor K (lm/W)
Green 555 nm	683
Red LED	60
Daylight without clouds	140
2850 K standard light source	16
2850 K standard light source with IR filter	350

Illuminance at imaging plane   Illuminance at a sensor imaging plane is described in this section [6]. Lux is generally used as an illumination unit. It is noted that lux is a photometric unit related to human eye characteristics, that is, it is not a pure physical unit. Illumination is defined as the light power per unit area. Suppose an optical system as shown in Fig. B.3. Here, a light flux F is incident on an object whose surface is taken to be completely diffusive. When light is reflected off an

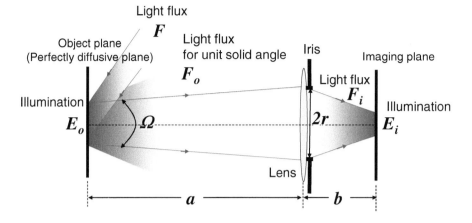

FIGURE B.3: Illumination at the imaging plane.

object with a perfect diffusion surface with reflectance R and area A, the reflected light uniformly diverges at the half whole solid angle π. Thus the light flux F_o divergent into a unit solid angle is calculated as

$$F_o = \frac{FR}{\pi}. \tag{B.2}$$

As the solid angle to the lens aperture or iris Ω is

$$\Omega = \frac{\pi r^2}{a^2}, \tag{B.3}$$

the light flux into the sensor imaging plane through the lens with a transmittance of T, F_i, is calculated as

$$F_i = F_o \Omega = FRT \left(\frac{r}{a}\right)^2 = FRT \frac{1}{4F_N^2} \left(\frac{m}{1+m}\right)^2. \tag{B.4}$$

If the lens has a magnification factor $m = b/a$, focal length f, and F-number $F_N = f/(2r)$, then

$$a = \frac{1+m}{m} f, \tag{B.5}$$

and thus

$$\left(\frac{r}{a}\right)^2 = \left(\frac{f}{2F_N}\right)^2 \left(\frac{m}{(1+m)f}\right)^2. \tag{B.6}$$

The illuminances at the object and at the sensor imaging plane are $E_o = F/A$ and $E_i = F_i/(m^2 A)$, respectively. It is noted that the object area is focused into the sensor imaging area multiplied by the square of the magnification factor m. By using the above equations, we obtain the following relation between the illuminance at an object plane E_o and that at the sensor imaging plane E_i:

$$E_i = \frac{E_o RT}{4F_N^2(1+m)^2} \cong \frac{E_o RT}{4F_N^2},$$

(B.7)

where $m \ll 1$ is used in the second equation, which is satisfied in a conventional imaging system. For example, E_i/E_o is about 1/30 when F_N is 2.8 and $T = R = 1$. It is noted that T and R are typically less than 1, so that this ratio is typically smaller than 1/30. It is noted that the illuminance at a sensor surface decreases to $1/10 - 1/100$ of the illuminance at an object.

C

Human eye and CMOS image sensors

In this chapter, we summarize the visual processing of human eyes, because they are an ideal imaging system and a model for CMOS imaging systems. To this end, we compare the human visual system with CMOS image sensors.

Retina

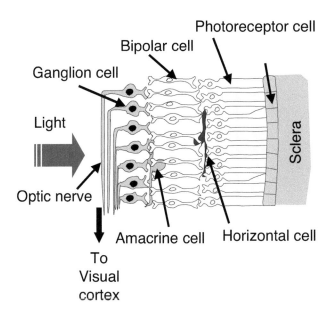

FIGURE C.1: Structure of the human retina [416].

The human eye has superior characteristics to state-of-the-art CMOS image sensors. The dynamic range of the human eye is about 200 dB with multi-resolution. In addition, the human eye has the focal plane processing function of spatio-temporal image preprocessing. In addition, humans have two eyes, which allows range finding by convergence and disparity. It is noted that distance measurements using disparity require complex processing at the visual cortex [621]. The human retina has an area of about 5 cm × 5 cm with a thickness of 0.4 mm [199,416,622,623]. The conceptual structure is illustrated in Fig. C.1. The incident light is detected by photoreceptors, which have two types, rod and cone.

Photosensitivity of the human eye

Rod photoreceptors have higher photosensitivity than cones and have adaptivity for light intensity, as shown in Fig. C.2 [624, 625]. Under uniform light illumination, the rod works in a range of two orders of magnitude with saturation characteristics. Figure C.2 schematically shows a photoresponse curve under constant illumination. The photoresponse curve adaptively shifts according to the environmental illumination and eventually converts over seven orders of magnitude. The human eye has a wide dynamic range under moonlight to sunlight due to this mechanism.

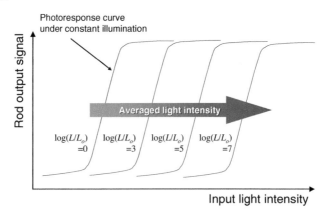

FIGURE C.2: Adaptation for light intensity in rod photoreceptors. The photoresponse curve shifts according to the average environmental illumination L. From the initial illumination L_o, the environmental illumination is changed exponentially in the range $\log(L/L_o) = 0$ to 7.

Color in retina

The human eye can sense color in a range of about 370 nm to 730 nm [623]. Rods are mainly distributed in the periphery of the retina and have a higher photosensitivity without color sensitivity, while cones are mainly concentrated in the center of the retina or fovea and have color sensitivity with less photosensitivity than rods. Thus the retina has two types of photoreceptors with high and low photosensitivities. When rods initiate vision, typically when the illumination is dark, the vision is called scotopic. When cones initiate vision, it is called photopic. The peak wavelengths for scotopic and photopic vision are 507 nm and 555 nm, respectively. For color sensitivity, rods are classified into L, M, and S types [626], which have similar characteristics of on-chip color filters in image sensors of R, G, and B, respectively. The center wavelengths of L, M, and S cones are 565 nm, 545 nm, and 440 nm, respectively [623]. The color sensitivity is different from that of animals; for example, some butterflies have a sensitivity in the ultraviolet range [301]. Surprisingly, the distribution of L, M, and S cones is not uniform [627], while image sensors have a regular arrangement of color filters, such as a Bayer pattern.

Comparison of human retina and a CMOS image sensor

Table C.1 summarizes a comparison of the human retina and a CMOS image sensor [199, 416, 622, 623].

TABLE C.1
Comparison of the human retina and a CMOS image sensor

Item	Retina	CMOS sensor
Resolution	Cones: 5×10^6 Rods: 10^8 Ganglion cells: 10^6	1–10×10^6
Size	Rod: diameter 1 μm near fovea Cone: diameter 1–4 μm in fovea, 4–10 μm in extrafovea	2–10 μm sq.
Color	3 Cones (L, M, S) (L+M): S = 14:1	On-chip RGB filter R:G:B=1:2:1
Minimum detectable illumination	$\sim$0.001 lux	0.1–1 lux
Dynamic range	Over 140 dB (adaptive)	60–70 dB
Detection method	Cis–trans isomerization $\rightarrow$two-stage amplification (500 $\times$ 500)	e–h pair generation Charge accumulation
Response time	$\sim$10 msec	Frame rate (video rate: 33 msec)
Output	Pulse frequency modulation	Analog voltage or digital
Number of outputs	Number of GCs: $\sim$1M	One analog output or bit-number in digital output
Functions	Photoelectronic conversion Adaptive function Spatio-temporal signal processing	Photoelectronic conversion Amplification Scanning

D

Wavelength region in visible and infrared lights

TABLE D.1

Wavelength region in visible and infrared lights

Wavelength region name	Wavelength	Photon energy
Ultra violet (UV)	200 – 380 nm	3.26 – 6.20 eV
Violet (V)	380 – 450 nm	2.76 – 3.26 eV
Blue (B)	450 – 495 nm	2.50 – 2.76 eV
Green (G)	495 – 570 nm	2.18 – 2.50 eV
Yellow (Y)	570 – 590 nm	2.10 – 2.18 eV
Orange (O)	590 – 620 nm	1.99 – 2.10 eV
Red (R)	620 – 750 nm	1.65 – 1.99 eV
Near infrared (NIR)	$0.75 - 1.4 \ \mu m$	0.9 –1.65 eV
Short-wavelength IR (SWIR)	$1.4 - 3 \ \mu m$	0.4 – 0.9 eV
Mid-wavelength IR (MWIR)	$3 - 8 \ \mu m$	0.15 – 0.4 eV
Long-wavelength IR (LWIR) (Thermal IR (TIR))	$8 - 15 \ \mu m$	80 – 150 meV
Far IR (FIR)	$15 - 1000 \ \mu m$	1.2– 80 meV

E

Fundamental characteristics of MOS capacitors

A MOS capacitor is composed of a metallic electrode (usually heavily doped poly-silicon is used) and a semiconductor with an insulator (usually SiO_2) in between. A MOS capacitor is an important part of a MOSFET and is easily implemented in standard CMOS technology by connecting the source and drain of a MOSFET. In this case, the gate and body of the MOSFET act as the electrodes of a capacitor. The characteristics of MOS capacitors are dominated by the channel underneath the insulator or SiO_2. It is noted that a MOS capacitor is a series sum of the two capacitors, a gate oxide capacitor C_{ox} and a depletion region capacitor C_D.

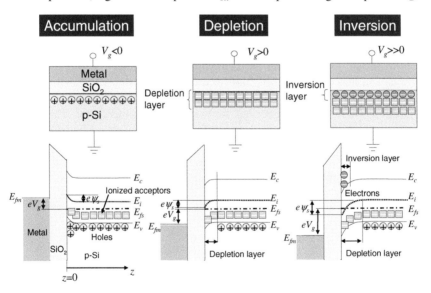

FIGURE E.1: Conceptual illustration of the three modes of MOS capacitor operation, accumulation, depletion, and inversion. E_c, E_v, E_{fs}, and E_i are the conduction band edge, valence band edge, Fermi energy of the semiconductor, and mid-gap energy, respectively. E_{fm} is the Fermi energy of the metal. V_g is the gate bias voltage.

There are three modes in a MOS capacitor, accumulation: depletion, and inversion, as shown in Fig. E.1, which are characterized by the surface potential ψ_s [79]. The surface potential $e\psi_s$ is defined as the difference of the mid-gap energy between the surface $(z = 0)$ and the bulk region $(z = \infty)$.

- $\psi_s < 0$: accumulation mode

- $\psi_B > \psi_s > 0$: depletion mode

- $\psi_s > \psi_B$: inversion mode

Here $e\psi_B$ is defined as the difference between the mid-gap energy at the bulk region $E_i(\infty)$ and the Fermi energy E_{fs}. In the accumulation mode, the gate bias voltage is negative, $V_g < 0$, and holes accumulate near the surface. This mode is rarely used in image sensors. In the depletion mode, a positive gate bias $V_g > 0$ is applied and causes free carriers to be depleted near the surface region. Space charges of ionized acceptors are located in the depletion region and compensate the induced charges by the gate voltage V_g. In this mode, the surface potential ψ_s is positive but smaller than ψ_B. The third mode is the inversion mode, which is used in a MOSFET when it turns on and in CMOS image sensors to accumulate photo-generated charges. To apply a larger gate bias voltage than in the depletion mode, an inversion layer appears, where electrons accumulate in the surface region. When E_i at $z = 0$ intersects E_{fs}, the inversion mode occurs, where $\psi_s = \psi_B$. If $psi_s > 2\psi_B$, then the surface is completely inverted, that is, it becomes an n-type region in this case. This mode is called strong inversion, while the $\psi_s < 2\psi_B$ mode is called weak inversion. It is noted that the electrons in the inversion layer are thermally generated electrons and/or diffusion electrons and hence it takes some time to establish inversion layers with electrons. This means that the inversion layer in the non-equilibrium state can act as a reservoir for electrons when they are generated, for example, by incident light. This reservoir is called a potential well for photo-generated carriers. If the source and drain regions are located on either side of the inversion layer, as in a MOSFET, electrons are quickly supplied to the inversion layer from the source and drain regions, so that an inversion layer filled with electrons is established in a very short time.

F

Fundamental characteristics of MOSFET

Enhancement and depletion types

MOSFETs are classified into two types: enhancement types and depletion types. Usually in CMOS sensors, enhancement types are used, although some sensors use depletion type MOSFETs. In the enhancement type, the threshold of an NMOSFET is positive, while in the depletion type the threshold of an NMOSFET is negative. Thus depletion type NMOSFETs can turn on without an applied gate voltage, that is, a normally ON state. In a pixel circuit, the threshold voltage is critical for operation, so that some sensors use depletion type MOSFET in the pixels [114].

Operation region

The operation of MOSFETs is first classified into two regions: above threshold and below threshold (sub-threshold). In each of these regions, three sub-regions exist, cutoff, linear (triode), and saturation. In the cutoff region, no drain current flows. Here we summarize the characteristics in each region for NMOSFET.

Above threshold: $V_{gs} > V_{th}$

Linear region   The condition of the linear region above threshold is

$$V_{gs} > V_{th},$$
$$V_{ds} < V_{gs} - V_{th}. \tag{F.1}$$

In the above condition, the drain current I_d is expressed as

$$I_d = \mu_n C_{ox} \frac{W_g}{L_g} \left[(V_{gs} - V_{th}) V_{ds} - \frac{1}{2} V_{ds}^2 \right], \tag{F.2}$$

where C_{ox} is the capacitance of the gate oxide per unit area and W_g and L_g are the gate width and length, respectively.

Saturation region

$$V_{gs} > V_{th},$$
$$V_{ds} > V_{gs} - V_{th}. \tag{F.3}$$

In the above condition,

$$I_d = \frac{1}{2} \mu_n C_{ox} \frac{W_g}{L_g} (V_{gs} - V_{th})^2. \tag{F.4}$$

For short channel transistors, channel length modulation effect must be considered, and thus Eq. F.4 is modified as [134]

$$I_d = \frac{1}{2} \mu_n C_{ox} \frac{W_g}{L_g} (V_{gs} - V_{th})^2 (1 + \lambda V_{ds})$$
$$= I_{sat}(V_{gs})(1 + \lambda V_{ds}), \tag{F.5}$$

where $I_{sat}(V_{gs})$ is the saturation drain current at V_{gs} without the channel length modulation effect. Equation F.5 means that even in the saturation region, the drain current gradually increases according to the drain–source voltage. In a bipolar transistor, a similar effect is called the Early effect and the characteristics parameter is the Early voltage V_E. In a MOSFET, the Early voltage V_E is $1/\lambda$ from Eq. F.5.

Sub-threshold region

In this region, the following condition is satisfied:

$$0 < V_{gs} < V_{th}. \tag{F.6}$$

In this condition, the drain current still flows and is expressed as [628]

$$I_d = I_o \exp\left[\frac{e}{m k_B T}\left(V_{gs} - V_{th} - \frac{m k_B T}{e}\right)\right]\left[1 - \exp\left(-\frac{e}{k_B T} V_{ds}\right)\right], \tag{F.7}$$

Here m is the body-effect coefficient [629], defined later, and I_o is given by

$$I_o = \mu_n C_{ox} \frac{W_g}{L_g} \frac{1}{m} \left(\frac{m k_B T}{e}\right)^2. \tag{F.8}$$

The intuitive method to extract the above sub-threshold current is given in Ref. [44, 630]. Some smart image sensors utilize the sub-threshold operation, and hence we briefly consider the origin of the sub-threshold current after the treatment in Refs. [44, 630].

The drain current in the sub-threshold region originates from the diffusion current, which is caused by differences of electron density between the source and drain ends, n_s and n_d, that is,

$$I_d = -q W_g x_c D_n \frac{n_d - n_s}{L_g}, \tag{F.9}$$

where x_c is the channel depth. It is noted that the electron density at each end $n_{s,d}$ is determined by the electron barrier height between each end and the flat channel $\Delta E_{s,d}$, and thus

$$n_{s,d} = n_o \exp\left(-\frac{\Delta E_{s,d}}{k_B T}\right), \tag{F.10}$$

where n_o is a constant. The energy barrier in each end is given by

$$\Delta E_{s,d} = -e\psi_s + e(V_{bi} + V_{s,d}), \tag{F.11}$$

where ψ_s is the surface potential at the gate. In the sub-threshold region, ψ_s is roughly a linear function of the gate voltage V_{gs} as

$$\psi_s = \psi_o + \frac{V_{gs}}{m}. \tag{F.12}$$

Here m is the body-effect coefficient given by

$$m = 1 + \frac{C_d}{C_{ox}}, \tag{F.13}$$

were C_d is the capacitance of the depletion layer per unit area. It is noted that $1/m$ is a measure of the capacitive coupling ratio from gate to channel. By using Eqs. F.9, F.10, F.11, and F.12, Eq. F.7 is obtained when the source voltage is connected to the ground voltage. The sub-threshold slope S is conventionally used for the measurement of the sub-threshold characteristics and defined as

$$S = \left(\frac{d(\log_{10} I_{ds})}{dV_{gs}}\right)^{-1} = 2.3 \frac{m k_B T}{e} = 2.3 \frac{k_B T}{e}\left(1 + \frac{C_d}{C_{ox}}\right) \tag{F.14}$$

The value of S is typically 70–100 mV/decade.

Linear region The sub-threshold region is also classified into linear and saturation regions, as well as the region above threshold. In the linear region, I_d depends on V_{ds}, while in the saturation region, I_d is almost independent of V_{ds}.

In the linear region, V_{ds} is small and both diffusion currents from the drain and source contribute to the drain current. Expanding Eq. F.7 under the condition $V_{ds} < k_B T/e$ gives

$$I_d = I_o \exp\left[\frac{e}{m k_B T}(V_{gs} - V_{th})\right]\frac{e}{k_B T} V_{ds}, \tag{F.15}$$

which shows that I_d is linear with V_{ds}.

Saturation region In this region, V_{ds} is larger than $k_B T/e$ and thus Eq. F.7 becomes

$$I_d = I_o \exp \left[\frac{e}{m k_B T} (V_{gs} - V_{th}) \right].$$

(F.16)

In this region, the drain current is independent of the drain–source voltage and only depends on the gate voltage when the source voltage is constant. The transition from the linear to the saturation region occurs around $V_{ds} \approx 4 k_B T/e$, which is about 100 mV at room temperature [630].

G

Optical format and resolution

TABLE G.1
Optical Format [631]

Format	Diagonal (mm)	H (mm)	V (mm)	Comment
$1/n$ inch	$16/n$	$12.8/n$	$9.6/n$	$n < 3$
$1/n$ inch	$18/n$	$14.4/n$	$10.8/n$	$n \geq 3$
35 mm	43.27	36.00	24.00	a.r.* 3:2
APS-C	27.26	22.7	15.1	
Four-thirds	21.63	17.3	13.0	a.r. 4:3

*aspect ratio

TABLE G.2
Resolution [2]

Abbreviation	Full name	Pixel number
CIF	Common Intermediate Format	352×288
QCIF	Quarter Common Intermediate Format	176×144
VGA	Video Graphics Array	640×480
QVGA	Quarter VGA	320×240
SVGA	Super VGA	800×400
XGA	eXtended Graphics Array	1024×768
UXGA	Ultra XGA	1600×1200
HD	High Definition	1280×720
2K (1080i), Full HD		1920×1080
4K (2160p)		3840×2160
8K (4320p), Super HD		7680×4320

H

Intrinsic optical signal and in vivo window

Intrinsic optical signal is an optical signal induced by biological activities such as optical absorption change according to the change of hemoglobin conformation. The change of hemoglobin conformation occurs according to oxidation and reduction reaction, which suggests that the brain activity is active. Figure H.1 shows absorption change of hemoglobin (Hb) and oxidized Hb (HbO$_2$) [632].

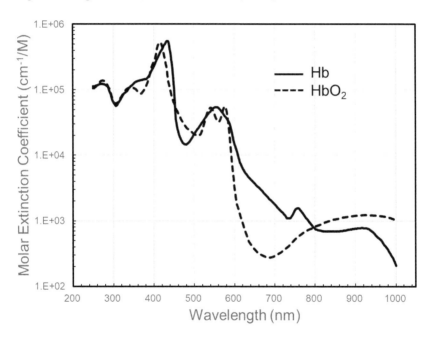

FIGURE H.1: Absorption spectrum of hemoglobin (Hb) and oxidized Hb (HbO$_2$). Illustrated after [632].

Figure H.2 shows an absorption coefficient of water and hemoglobin as a function of wavelength [633]. The region where both absorption of water and hemoglobin is low is known as "*in vivo* window". This region is useful when light is required to be introduced inside the body due to low absorption.

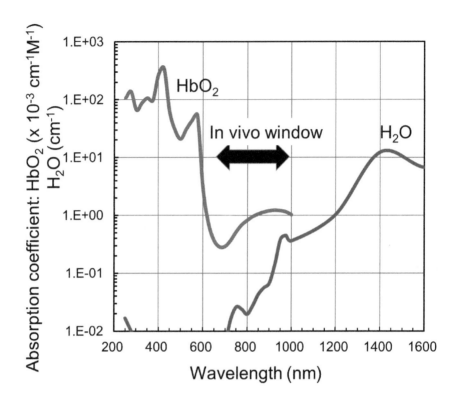

FIGURE H.2: The absorption coefficient of water [634] and hemoglobin [632] as a function of wavelength [633]. The region where both absorptions are low is called an in vivo window.

References

[1] A. Moini. *Vision Chips*. Kluwer Acadmic Publisher, Dordrecht, The Netherlands, 2000.

[2] J. Nakamura, editor. *Image Sensors and Signal Processing for Digital Still Cameras*. CRC Press, Boca Raton, FL, 2005.

[3] K. Yonemoto. *Fundamentals and Applications of CCD/CMOS Image Sensors*. CQ Pub. Co.,Ltd., Tokyo, Japan, 2003. In Japanese.

[4] T. Kuroda. *Essential Principles of Image Sensors*. CRC Press, Boca Raton, FL, USA, 2015.

[5] Y. Takemura. *CCD Camera Technologies*. Corona Pub., Co. Ltd., Tokyo, Japan, 1997. In Japanese.

[6] Y. Kiuchi. *Fundamentals and Applications of Image Sensors*. The Nikkankogyo Shimbun Ltd., Tokyo, Japan, 1991. In Japanese.

[7] T. Ando and H. Komobuchi. *Introduction of Solid-State Image Sensors*. Nihon Riko Syuppan Kai, Tokyo, Japan, 1999. In Japanese.

[8] K. Aizawa and T. Hamamoto, editors. *CMOS Image Sensors*. Corona Pub., Co. Ltd., Tokyo, Japan, 2012. In Japanese.

[9] A.J.P. Theuwissen. *Solid-State Imaging with Charge-Coupled Devices*. Kluwer Academic Pub., Dordrecht, The Netherlands, 1995.

[10] Y. Takemura. The Development of Video-Camera Technologies: Many Innovations Behind Video Cameras Are Used for Digital Cameras and Smartphones. *IEEE Consumer Electronics Magazine*, 8(4):10–16, July 2019.

[11] S. Morrison. A new type of photosensitive junction device. *Solid-State Electron.*, 5:485–494, 1963.

[12] J.W. Horton, R.V. Mazza, and H. Dym. The scanistor – A solid-state image scanner. *Proc. IEEE*, 52(12):1513 – 1528, December 1964.

[13] P.K. Weimer, G. Sadasiv, Jr. J.E. Meyer, L. Meray-Horvath, and W.S. Pike. A self-scanned solid-state image sensor. *Proc. IEEE*, 55(9), 1967.

[14] M.A. Schuster and G. Strull. A monolithic mosaic of photon sensors for solid-state imaging applications. *IEEE Trans. Electron Dev.*, ED-13(12):907 – 912, December 1966.

[15] G.P. Weckler. Operation of p-n Junction Photodetectors in a Photon Flux Integration Mode. *IEEE J. Solid-State Circuits*, SC-2(3):65–73, September 1967.

[16] R.H. Dyck and G.P. Weckler. Integrated arrays of silicon photodetectors for image sensing. *IEEE Trans. Electron Dev.*, ED-15(4):196–201, April 1968.

[17] P.J. Noble. Self-scanned silicon image detector arrays. *IEEE Trans. Electron Dev.*, ED-15(4):202–209, April 1968.

[18] R.A. Anders, D.E. Callahan, W.F. List, D.H. McCann, and M.A. Schuster. Developmental solid-state imaging system. *IEEE Trans. Electron Dev.*, ED-15(4):191– 196, April 1968.

[19] G. Sadasiv, P.K. Weimer, and W.S. Pike. Thin-film circuits for scanning image-sensor arrays. *IEEE Trans. Electron Dev.*, ED-15(4):215– 219, April 1968.

[20] W.F. List. Solid-state imaging – Methods of approach. *IEEE Trans. Electron Dev.*, ED-15(4):256–261, April 1968.

[21] W.S. Boyle and G.E. Smith. Charge-Coupled Semiconductor Devices. *Bell System Tech. J.*, 49:587–593, 1970.

[22] G.F. Amelio, M.F. Tompsett, and G.E. Smith. Experimental Verification of the Charge-Coupled Semiconductor Device Concept. *Bell System Tech. J.*, 49:593–600, 1970.

[23] M.F. Tompsett. Charge Transfer Imaging Devices, April 18 1978. US Patent 4,085,456.

[24] G.E. Smith. The invention of the CCD. *Nuclear Instruments & Methods in Physics Research A*, 2001.

[25] N. Koike, I. Takemoto, K. Satoh, S. Hanamura, S. Nagahara, and M. Kubo. MOS Area Sensor: Part I – Design Consideration and Performance of an n-p-n Structure 484 × 384 Element Color MOS Imager. *IEEE Trans. Electron Dev.*, 27(8):1682 – 1687, August 1980.

[26] H. Nabeyama, S. Nagahara, H. Shimizu, M. Noda, and M. Masuda. All Solid State Color Camera With Single-Chip Mos Imager. *IEEE Trans. Consumer Electron.*, CE-27(1):40 – 46, February 1981.

[27] S. Ohba, M. Nakai, H. Ando, S. Hanamura, S. Shimada, K. Satoh, K. Takahashi, M. Kubo, and T. Fujita. MOS Area Sensor: Part II – Low-Noise MOS Area Sensor with Antiblooming Photodiodes. *IEEE J. Solid-State Circuits*, 15(4):747 – 752, August 1980.

[28] M. Aoki, H. Ando, S. Ohba, I. Takemoto, S. Nagahara, T. Nakano, M. Kubo, and T. Fujita. 2/3-Inch Format MOS Single-Chip Color Imager. *IEEE J. Solid-State Circuits*, 17(2):375– 380, April 1982.

[29] H. Ando, S. Ohba, M. Nakai, T. Ozaki, N. Ozawa, K. Ikeda, T. Masuhara, T. Imaide, I. Takemoto, T. Suzuki, and T. Fujita. Design consideration and performance of a new MOS imaging device. *IEEE Trans. Electron Dev.*, ED-32(8):1484– 1489, August 1985.

[30] T. Kinugasa, M. Noda, T. Imaide, I. Aizawa, Y. Todaka, and M Ozawa. An Electronic Variable-Shutter System in Video Camera Use. *IEEE Trans. Consumer Electron.*, CE-33(3):249 – 255, August 1987.

[31] T. Nakamura, K. Matsumoto, R. Hyuga, and A Yusa. A new MOS image sensor operating in a non-destructive readout mode. In *Tech. Dig. Int'l Electron Devices Meeting (IEDM)*, pages 353 – 356, 1986.

[32] J. Hynecek. A new device architecture suitable for high-resolution and high-performance image sensors . *IEEE Trans. Electron Dev.*, 35(5):646–652, May 1988.

[33] N. Tanaka, T. Ohmi, and Y. Nakamura. A novel bipolar imaging device with self-noise-reduction capability. *IEEE Trans. Electron Dev.*, 36(1):31 – 38, January 1989.

[34] A. Yusa, J. Nishizawa, M. Imai, H. Yamada, J. Nakamura, T. Mizoguchi, Y. Ohta, and M. Takayama. SIT Image sensor: Design considerations and characteristics. *IEEE Trans. Electron Dev.*, 33(6):735 – 742, June 1986.

[35] F. Andoh, K. Taketoshi, K. Nakamura, and M. Imai. 'AMI' A New Amplified Solid State Imager. *J. Inst. Image Information & Television Eng.*, 41(11):1075–1082, November 1987. In Japanese.

[36] F. Andoh, K. Taketoshi, J. Yamazaki, M. Sugawara, Y. Fujita, K. Mitani, Y. Matuzawa, K. Miyata, and S. Araki. A 250000-pixel image sensor with FET amplification at each pixel for high-speed television cameras. *Dig. Tech. Papers Int'l Solid-State Circuits Conf. (ISSCC)*, pages 212 – 213, 298, February 1990.

[37] M. Kyomasu. A New MOS Imager Using Photodiode as Current Source. *IEEE J. Solid-State Circuits*, 26(8):1116–1122, August 1991.

[38] M. Sugawara, H. Kawashima, F. Andoh, N. Murata, Y. Fujita, and M. Yamawaki. An amplified MOS imager suited for image processing. *Dig. Tech. Papers Int'l Solid-State Circuits Conf. (ISSCC)*, pages 228 – 229, February 1994.

[39] M. Yamawaki, H. Kawashima, N. Murata, F. Andoh, M. Sugawara, and Y. Fujita. A pixel size shrinkage of amplified MOS imager with two-line mixing. *IEEE Trans. Electron Dev.*, 1996.

[40] E. R. Fossum. Active pixel sensors – Are CCD's dinosaurs? In *Proc. SPIE*, volume 1900 of *Charge-Coupled Devices and Optical Sensors III*, pages 2–14, 1993.

[41] E.R. Fossum. CMOS Image Sensors: Electronic Camera-On-A-Chip. *IEEE Trans. Electron Dev.*, 44(10):1689–1698, October 1997.

[42] R.M. Guidash, T.-H. Lee, P.P.K. Lee, D.H. Sackett, C.I. Drowley, M.S. Swenson, L. Arbaugh, R. Hollstein, F. Shapiro, and S. Domer. A 0.6 μm CMOS pinned photodiode color imager technology. In *Tech. Dig. Int'l Electron Devices Meeting (IEDM)*, pages 927 – 929, December 1997.

[43] H. S. Wong. Technology and device scaling considerations for CMOS imagers. *IEEE Trans. Electron Dev.*, 43(12):2131–2141, December 1996.

[44] C. Mead. *Analog VLSI and Neural Systems*. Addison-Wesley Publishing Company, Reading, MA, 1989.

[45] C. Koch and H. Liu. *VISION CHIPS, Implementing Vision Algorithms with Analog VLSI Circuits*. IEEE Computer Society, Los Alamitos, CA, 1995.

[46] T.M. Bernard, B.Y. Zavidovique, and F.J. Devos. A Programmable Artificial Retina. *IEEE J. Solid-State Circuits*, 28(7):789–798, July 1993.

[47] S. Kameda and T. Yagi. An analog VLSI chip emulating sustained and transient response channels of the vertebrate retina. *IEEE Trans. Neural Networks*, 14(5):1405 – 1412, September 2003.

[48] R. Takami, K. Shimonomura, S. Kamedaa, and T. Yagi. An image pre-processing system employing neuromorphic 100 × 100 pixel silicon retina. In *Int'l Symp. Circuits & Systems (ISCAS)*, pages 2771–2774, Kobe, Japan, May 2005.

[49] http://www.itrs.net/.

[50] G. Moore. Cramming more components onto integrated circuits. *Electronics*, 38(8), April 1965.

[51] L. Fortuna, P. Arena, D. Balya, and A. Zarandy. Cellular neural networks: a paradigm for nonlinear spatio-temporal processing. *IEEE Circuits & Systems Mag.*, 1(4):6–21, 2001.

[52] S. Espejo, A. Rodríguez-Vázquez, R. Domínguez-Castro, J.L. Huertas, and E. Sánchez-Sinencio. Smart-pixel cellular neural networks in analog current-mode CMOS technology. *IEEE J. Solid-State Circuits*, 29(8):895 – 905, August 1994.

[53] R. Domínguez-Castro, S. Espejo, A. Rodríguez-Vázquez, R.A. Carmona, P. Földesy, A. Zárandy, P. Szolgay, T. Szirányi, and T. Roska. A 0.8-μm CMOS two-dimensional programmable mixed-signal focal-plane array processor with on-chip binary imaging and instructions storage. *IEEE J. Solid-State Circuits*, 32(7):1013 – 1026, July 1997.

[54] G. Liñan-Cembrano, L. Canranza, S. Espejo, R. Dominguez-Castro, and A. Rodíguez-Vázquez. CMOS mixed-signal flexible vision chips. In M. Valle,

editor, *Smart Adaptive Systems on Silicon*, pages 103–118. Kluwer Academic Pub., Dordecht, The Netherlands, 2004.

[55] R. Etienne-Cummings, Z.K. Kalayjian, and D. Cai. A Programmable Focal-Plane MIMD Image Processor Chip. *IEEE J. Solid-State Circuits*, 36(1):64–73, January 2001.

[56] E. Culurciello, R. Etienne-Cummings, and K.A. Boahen. A Biomorphic Digital Image Sensor. *IEEE J. Solid-State Circuits*, 38(2):281–294, February 2003.

[57] P. Dudek and P.J. Hicks. A General-Purpose Processor-per-Pixel Analog SIMD Vision Chip. *IEEE Trans. Circuits & Systems I*, 52(1):13–20, January 2005.

[58] P. Dudek and P.J. Hicks. A CMOS General-Purpose Sampled-Data Analog Processing Element. *IEEE Trans. Circuits & Systems II*, 47(5):467–473, May 2000.

[59] M. Ishikawa, K. Ogawa, T. Komuro, and I. Ishii. A CMOS vision chip with SIMD processing element array for 1 ms image processing. In *Dig. Tech. Papers Int'l Solid-State Circuits Conf. (ISSCC)*, pages 206–207, February 1999.

[60] M. Ishikawa and T. Komuro. Digital vision chips and high-speed vision systems. In *Dig. Tech. Papers Symp. VLSI Circuits*, pages 1–4, June 2001.

[61] N. Mukohzaka, H. Toyoda, S. Mizuno, M.H. Wu, Y. Nakabo, and M. Ishikawa. Column parallel vision system: CPV. In *Proc. SPIE*, volume 4669, pages 21–28, San Jose, CA, January 2002.

[62] T. Komuro, S. Kagami, and M. Ishikawa. A high soeed digital vision chip with multi-grained parallel processing capability. In *IEEE Workshop on Charge-Coupled Devices & Advanced Image Sensors*, Elmau, Geramany, June 2003.

[63] T. Komuro, I. Ishii, M. Ishikawa, and A. Yoshida. A Digital Vision Chip Specialized for High-Speed Target Tracking. *IEEE Trans. Electron Dev.*, 50(1):191–199, January 2003.

[64] T. Komuro, S. Kagami, and M. Ishikawa. A dynamically reconfigurable SIMD processor for a vision chip. *IEEE J. Solid-State Circuits*, 39(1):265–268, January 2004.

[65] T. Komuro, S. Kagami, M. Ishikawa, and Y. Katayama. Development of a Bit-level Compiler for Massively Parallel Vision Chips. In *IEEE Int'l Workshop Computer Architecture for Machine Perception (CAMP)*, pages 204–209, Palermo, July 2005.

[66] D. Renshaw, P.B. Denyer, G. Wang, and M. Lu. ASIC VISION. In *Proc. Custom Integrated Circuits Conf. (CICC)*, pages 7.3/1–7.3/4, May 1990.

[67] P.B. Denyer, D. Renshaw, Wang G., and Lu M. CMOS image sensors for multimedia applications. In *Proc. Custom Integrated Circuits Conf. (CICC)*, pages 11.5.1 – 11.5.4, May 1993.

[68] K. Chen, M. Afghani, P.E. Danielsson, and C. Svensson. PASIC: A processor-A/D converter-sensor integrated circuit. In *Int'l Symp. Circuits & Systems (ISCAS)*, volume 3, pages 1705 – 1708, May 1990.

[69] J.-E. Eklund, C. Svensson, and A. Åström. VLSI Implementation of a Focal Plane Image Processor – A Realization of the Near-Sensor Image Processing Concept. *IEEE Trans. VLSI Systems*, 4(3):322–335, September 1996.

[70] L. Lindgren, J. Melander, R. Johansson, and B. Moller. A multiresolution 100-GOPS 4-Gpixels/s programmable smart vision sensor for multisense imaging. *IEEE J. Solid-State Circuits*, 40(6):1350–1359, June 2005.

[71] E.D. Palik. *Handbook of Optical Constants of Solids*. Academic Press, New York, NY, 1977.

[72] S.E. Swirhun, H.-H. Kwark, and R.M. Swanson. Measurement of electron lifetime, electron mobility and band-gap narrowing in heavily doped p-type silicon. In *Tech. Dig. Int'l Electron Devices Meeting (IEDM)*, pages 24–27, 1986.

[73] J. del Alamo, S. Swirhun, and R.M. Swanson. Measuring and modeling minority carrier transport in heavily doped silicon. *Solid-State Electronic.*, 28:47–54, 1985.

[74] J. del Alamo, S. Swirhun, and R.M. Swanson. Simultaneous measurement of hole lieftime, hole mobility and bandgap narrowing in heavily doped n-type silicon. In *Tech. Dig. Int'l Electron Devices Meeting (IEDM)*, pages 290–293, 1985.

[75] J.S. Lee, R.I. Hornsey, and D. Renshaw. Analysis of CMOS Photodiodes. I. Quantum efficiency. *IEEE Trans. Electron Dev.*, 50(5):1233 – 1238, May 2003.

[76] J.S. Lee, R.I. Hornsey, and D. Renshaw. Analysis of CMOS Photodiodes. II. Lateral photoresponse. *IEEE Trans. Electron Dev.*, 50(5):1239 – 1245, May 2003.

[77] S.G. Chamberlain, D.J. Roulston, and S.P. Desai. Spectral Response Limitation Mechanisms of a Shallow Junction n+-p Photodiode. *IEEE J. Solid-State Circuits*, SC-13(1):167– 172, February 1978.

[78] I. Murakami, T. Nakano, K. Hatano, Y. Nakashiba, M. Furumiya, T. Nagata, H. Utsumi, S. Uchida, K. Arai, N. Mutoh, A. Kohno, N. Teranishi, and Y. Hokari. New Technologies of Photo-Sensitivity Improvement and VOD Shutter Voltage Reduction for CCD Image Sensors. In *Proc. SPIE*, volume 3649, pages 14–21, San Jose, CA, 1999.

[79] S.M. Sze. *Physics of Semiconductor Devices*. John Wiley & Sons, Inc., New York, NY, 1981.

[80] N.V. Loukianova, H.O. Folkerts, J.P.V. Maas, D.W.E. Verbugt, A.J. Mierop, W. Hoekstra, E. Roks, and A.J.P. Theuwissen. Leakage current modeling of test structures for characterization of dark current in CMOS image sensors. *IEEE Trans. Electron Dev.*, 50(1):77–83, January 2003.

[81] B. Pain, T. Cunningham, B. Hancock, C. Wrigley, and C. Sun. Excess Noise and Dark Current Mechanism in CMOS Imagers. In *IEEE Workshop on Charge-Coupled Devices & Advanced Image Sensors*, pages 145–148, Karuizawa, Japan, June 2005.

[82] A.S. Grove. *Physics and Technology of Semiconductor Devices*. John Wiley & Sons, Inc., New York, NY, 1967.

[83] G.A.M. Hurkx, H.C. de Graaff, W.J. Kloosterman, and M.P.G. Knuvers. A new analytical diode model including tunneling and avalanche breakdown. *IEEE Trans. Electron Dev.*, 39(9):2090–2098, September 1992.

[84] H.O. Folkerts, J.P.V. Maas, D.W.E. Vergugt, A.J.Mierop, W. Hoekstra, N.V. Loukianova, and E. Rocks. Characterization of Dark Current in CMOS Image Sensors. In *IEEE Workshop on Charge-Coupled Devices & Advanced Image Sensors*, Elmau, Germany, May 2003.

[85] H. Zimmermann. *Silicon Optoelectronic Integrated Cicuits*. Springer-Verlag, Berlin, Germany, 2004.

[86] S. Radovanović, A.-J. Annema, and B. Nauta. *High-Speed Photodiodes in Standard CMOS Technology*. Springer, Dordecht, The Netherlands, 2006.

[87] B. Razavi. *Design of Integrated Circuits for Optical Communications*. McGraw-Hill Companies Inc., New York, NY, 2003.

[88] J. Singh. *Semiconductor Optoelectronics: Physics and Technology*. McGraw-Hill, Inc., New York, NY, 1995.

[89] V. Brajovic, K. Mori, and N. Jankovic. 100 frames/s CMOS Range Image Sensor. In *Dig. Tech. Papers Int'l Solid-State Circuits Conf. (ISSCC)*, pages 256 – 257, February 2001.

[90] Y. Takiguchi, H. Maruyama, M. Kosugi, F. Andoh, T. Kato, K. Tanioka, J. Yamazaki, K. Tsuji, and T. Kawamura. A CMOS Imager Hybridized to an Avalanche Multiplied Film. *IEEE Trans. Electron Dev.*, 44(10):1783–1788, October 1997.

[91] A. Biber, P. Seitz, and H. Jäckel. Avalanche Photodiode Image Sensor in Standard BiCMOS Technology. *IEEE Trans. Electron Dev.*, 47(11):2241–2243, November 2000.

[92] C. Niclass, A. Rochas, P.-A. Besse, and E. Charbon. Toward a 3-D Camera Based on Single Photon Avalanche Diodes. *IEEE Selected Topic Quantum Electron.*, 2004.

[93] C. Niclass, A. Rochas, P.-A. Besse, and E. Charbon. Design and Characterization of a CMOS 3-D Image Sensor Based on Single Photon Avalanche Diodes. *IEEE J. Solid-State Circuits*, 40(9):1847 – 1854, September 2005.

[94] C.J. Stapels, W.G. Lawrence, F.L. Augustine, and J.F. Christian. Characterization of a CMOS Geiger Photodiode Pixel. *IEEE Trans. Electron Dev.*, 53(4):631–635, April 2006.

[95] N.A.W. Dutton, L. Parmesan, A.J. Holmes, L.A. Grant, and R.K. Henderson. 320x240 Oversampled Digital Single Photon Counting Image Sensor. In *Dig. Tech. Papers Symp. VLSI Circuits*, Honolulu, HI, USA, June 2014.

[96] F. Zappa, S. Tisa, A. Tosi, and S. Cova. Principles and features of single-photon avalanche diode arrays. *Sensors & Actuators A*, 140:103–112, June 2007.

[97] H. Finkelstein, M.J. Hsu, and S.C. Esener. STI-Bounded Single-Photon Avalanche Diode in a Deep-Submicrometer CMOS Technology. *IEEE Electron Device Lett.*, 27(11):887–889, November 2006.

[98] M. Mori, Y. Sakata, M. Usuda, S. Yamahira, S. Kasuga, Y. Hirose, Y. Kato, and T. Tanaka. A 1280×720 Single-Photon-Detecting Image Sensor with 100dB Dynamic Range Using a Sensitivity-Boosting Technique. In *Dig. Tech. Papers Int'l Solid-State Circuits Conf. (ISSCC)*, San Francisco, CA, USA, February 2016.

[99] Y. Hirose, S. Koyama, M. Ishiiand, S. Saitou, M. Takemoto, Y. Nose, A. Inoue, Y. Sakata, Y. Sugiura, T. Kabe, M. Usuda, S. Kasuga, M. Mori, A. Odagawa, and T. Tanaka. A 250 m Direct Time-of-Flight Ranging System Based on a Synthesis of Sub-Ranging Images and a Vertical Avalanche Photo-Diodes (VAPD) CMOS Image Sensor. *Sensors*, 18:3642, 2018.

[100] M. Kubota, T. Kato, S. Suzuki, H. Maruyama, K. Shidara, K. Tanioka, K. Sameshima, T. Makishima, K. Tsuji, , T. Hirai, and T. Yoshida. Ultra high-sensitivity new super-HARP camera. *IEEE Trans. Broadcast.*, 42(3):251–258, September 1996.

[101] T. Watabe, M. Goto, H. Ohtake, H. Maruyama, M. Abe, K. Tanioka, and N. Egami. New signal readout method for ultrahigh-sensitivity CMOS image sensor. *IEEE Trans. Electron Dev.*, 50(1):63 – 69, January 2003.

[102] S. Aihara, Y. Hirano, T. Tajima, K. Tanioka, M. Abe, N. Saito, N. Kamata, and D. Terunuma. Wavelength selectivities of organic photoconductive films: Dye-doped polysilanes and zinc phthalocyanine/tris-8-hydroxyquinoline aluminum double layer. *Appl. Phys. Lett.*, 82(4):511–513, January 2003.

[103] T. Watanabe, S. Aihara, N. Egami, M. Kubota, K. Tanioka, N. Kamata, and D. Terunuma. CMOS Image Sensor Overlaid with an Organic Photoconductive Film. In *IEEE Workshop on Charge-Coupled Devices & Advanced Image Sensors*, pages 48–51, Karuizawa, Japan, June 2005.

[104] S. Takada, M. Ihama, and M. Inuiya. CMOS Image Sensor with Organic Photoconductive Layer Having Narrow Absorption Band and Proposal of Stack Type Solid-State Image Sensors. In *Proc. SPIE*, volume 6068, pages 60680A–1 – A8, San Jose, CA, January 2006.

[105] J. Burm, K.I. Litvin, D.W. Woodard, W.J. Schaff, P. Mandeville, M.A. Jaspan, M.M. Gitin, and L.F. Eastman. High-frequency, high-efficiency MSM photodetectors. *IEEE J. Quantum Electron.*, 31(8):1504 – 1509, August 1995.

[106] K. Tanaka, F. Ando, K. Taketoshi, I. Ohishi, and G. Asari. Novel Digital Photosensor Cell in GaAs IC Using Conversion of Light Intensity to Pulse Frequency. *Jpn. J. Appl. Phys.*, 32(11A):5002–5007, November 1993.

[107] E. Lange, E. Funatsu, J. Ohta, and K. Kyuma. Direct image processing using arrays of variable-sensitivity photodetectors. In *Dig. Tech. Papers Int'l Solid-State Circuits Conf. (ISSCC)*, pages 228 – 229, February 1995.

[108] H.-B. Lee, H.-S. An, H.-I. Cho, J.-H. Lee, and S.-H. Hahm. UV photo-responsive characteristics of an n-channel GaN Schottky-barrier MISFET for UV image sensors. *IEEE Electron Device Lett.*, 27(8):656 – 658, August 2006.

[109] M. Abe. Image sensors and circuit technologies. In T. Enomoto, editor, *Video/Image LSI System Desing Technology*, pages 208 – 248. Corona Pub., Co. Ltd., Tokyo, 2003. In Japanese.

[110] N. Teranishi, A. Kohono, Y. Ishihara, E. Oda, and K. Arai. No image lag photodiode structure in the interline CCD image sensor. In *Tech. Dig. Int'l Electron Devices Meeting (IEDM)*, pages 324– 327, 1982.

[111] B.C. Burkey, W.C. Chang, J. Littlehale, T.H. Lee, T.J. Tredwell, J.P. Lavine, and E.A. Trabka. The pinned photodiode for an interline-transfer CCD image sensor. In *Tech. Dig. Int'l Electron Devices Meeting (IEDM)*, pages 28– 31, 1984.

[112] I. Inouc, H. Ihara, H. Yamashita, T. Yamaguchi, H. Nozaki, and R. Miyagawa. Low dark current pinned photo-diode for CMOS image sensor. In *IEEE Workshop on Charge-Coupled Devices & Advanced Image Sensors*, pages 25–32, Karuizawa, Japan, June 1999.

[113] S. Agwani, R Cichomski, M. Gorder, A. Niederkorn, Sknow M., and K. Wanda. A 1/3" VGA CMOS Imaging System On a Chip. In *IEEE Workshop on Charge-Coupled Devices & Advanced Image Sensors*, pages 21–24, Karuizawa, Japan, June 1999.

[114] K. Yonemoto and H. Sumi.   A CMOS image sensor with a simple fixed-pattern-noise-reduction technology and a hole accumulation diode. *IEEE J. Solid-State Circuits*, 2000.

[115] E.R. Fossum and D.B. Hondongwa. A Review of the Pinned Photodiode for CCD and CMOS Image Sensors. *IEEE J. Electron Dev. Soc.*, 2(3):33 – 43, May 2014.

[116] A. Pelamatti, V. Goiffon, A. De Ipanema, P. Magnan, C. Virmontois, O. Saint-Peé, and M. Breart de Boisanger.   Absolute Pinning Voltage Measurement: Comparison between In-pixel and JFET Extraction Methods. In *Int'l Image Sensor Workshop*, Vaals, The Netherlands, June 2015.

[117] V. Goiffon, M. Estribeau, J. Michelot, P. Cervantes, A. Pelamatti, O. Marcelot, and P. Magnan.   Pixel Level Characterization of Pinned Photodiode and Transfer Gate Physical Parameters in CMOS Image Sensors. *IEEE J. Electron Dev. Soc.*, 2(4):65 – 76, July 2014.

[118] M. Noda, T. Imaide, T. Kinugasa, and R. Nishimura.   A Solid State Color Video Camera with a Horizontal Readout MOS Imager.   *IEEE Trans. Consumer Electron.*, CE-32(3):329 – 336, August 1986.

[119] S. Miyatake, M. Miyamoto, K. Ishida, T. Morimoto, Y. Masaki, and H. Tanabe. Transversal-readout architecture for CMOS active pixel image sensors. *IEEE Trans. Electron Dev.*, 50(1):121 – 129, January 2003.

[120] J.D. Plummer and J.D. Meindl. MOS electronics for a portable reading aid for the blind. *IEEE J. Solid-State Circuits*, 7(2):111 – 119, April 1972.

[121] L.J. Kozlowski, J. Luo, W.E. Kleinhans, and T. Liu. Comparison of Passive and Active Pixel Schemes for CMOS Visible Imagers. In *Proc. SPIE*, volume 3360 of *Infrared Readout Electronics IV*, pages 101–110, Orland, FL, April 1998.

[122] B. Pain, G. Yang, M. Ortiz, C. Wrigley, B. Hancock, and T. Cunningham. Analysis and enhancement of low-light-level performance of photodiode-type CMOS active pixel imagers operated with sub-threshold reset.   In *IEEE Workshop on Charge-Coupled Devices & Advanced Image Sensors*, pages 140–143, Karuizawa, Japan, June 1999.

[123] Y. Endo, Y. Nitta, H. Kubo, T. Murao, K. Shimomura, M. Kimura, K. Watanabe, and S. Komori. 4-micron pixel CMOS iamge sensor with low image lag and high-temperature operability. In *Proc. SPIE*, volume 5017, pages 196–204, Santa Clara, CA, January 2003.

[124] I. Inoue, N. Tanaka, H. Yamashita, T. Yamaguchi, H. Ishiwata, and H. Ihara. Low-leakage-current and low-operating-voltage buried photodiode for a CMOS imager. *IEEE Trans. Electron Dev.*, 50(1):43–47, January 2003.

[125] H. Takahashi, M. Kinoshita, K. Morita, T. Shirai, T. Sato, T. Kimura, H. Yuzurihara, and S. Inoue. A 3.9 μm pixel pitch VGA format 10 b digital

image sensor with 1.5-transistor/pixel. In *Dig. Tech. Papers Int'l Solid-State Circuits Conf. (ISSCC)*, pages 108–109, February 2004.

[126] M. Mori, M. Katsuno, S. Kasuga, T. Murata, and T. Yamaguchi. A 1/4in 2M pixel CMOS image sensor with 1.75 transistor/pixel. In *Dig. Tech. Papers Int'l Solid-State Circuits Conf. (ISSCC)*, pages 110–111, February 2004.

[127] K. Mabuchi, N. Nakamura, E. Funatsu, T. Abe, T. Umeda, T. Hoshino, R. Suzuki, and H. Sumi. CMOS image sensor using a floating diffusion driving buried photodiode. In *Dig. Tech. Papers Int'l Solid-State Circuits Conf. (ISSCC)*, pages 112–113, February 2004.

[128] R.D. McGrath, H. Fujita, R.M. Guidash, T.J. Kenney, and W. Xu. Shared pixels for CMOS image sensor arrays. In *IEEE Workshop on Charge-Coupled Devices & Advanced Image Sensors*, pages 9–12, Karuizawa, Japan, June 2005.

[129] M. Murakami, M. Masuyama, S. Tanaka, M. Uchida, K. Fujiwara, M Kojima, Y. Matsunaga, and S. Mayumi. 2.8 μm-Pixel Image Sensor νMaicovicon^{TM}. In *IEEE Workshop on Charge-Coupled Devices & Advanced Image Sensors*, pages 13–14, Karuizawa, Japan, June 2005.

[130] Y.C. Kim, Y.T. Kim, S.H. Choi, H.K. Kong, S.I. Hwang, J.H. Ko, B.S. Kim, T. Asaba, S.H. Lim, J.S. Hahn, J.H. Im, T.S. Oh, D.M. Yi, J.M. Lee, W.P. Yang, J.C. Ahn, E.S. Jung, and Y.H. Lee. 1/2-inch 7.2M Pixel CMOS Image Sensor with 2.25 μm Pixels Using 4-Shared Pixel Structure for Pixel-Level Summation. In *Dig. Tech. Papers Int'l Solid-State Circuits Conf. (ISSCC)*, pages 1994–2003, February 2006.

[131] S. Yoshihara, M. Kikuchi, Y. Ito, Y. Inada, S. Kuramochi, H. Wakabayashi, M. Okano, K. Koseki, H. Kuriyama, J. Inutsuka, A. Tajima, T. Nakajima, Y. Kudoh, F. Koga, Y. Kasagi, S. Watanabe, and T. Nomoto T. A 1/1.8-inch 6.4M Pixel 60 frames/s CMOS Image Sensor with Seamless Mode Change. In *Dig. Tech. Papers Int'l Solid-State Circuits Conf. (ISSCC)*, pages 1984 – 1993, February 2006.

[132] O. Yadid-Pecht, B. Pain, C. Staller, C. Clark, and E. Fossum. CMOS active pixel sensor star tracker with regional electronic shutter. *IEEE J. Solid-State Circuits*, 32(2):285–288, February 1997.

[133] S.E. Kemeny, R. Panicacci, B. Pain, L. Matthies, and E.R. Fossum. Multiresolution Image Sensor. *IEEE Trans. Circuits & Systems Video Tech.*, 7(4):575–583, August 1997.

[134] B. Razavi. *Design of Analog CMOS Integrated Circuits*. McGraw-Hill Companies, Inc., New York, NY, 2001.

[135] K. Salama and A. El Gamal. Analysis of active pixel sensor readout circuit. *IEEE Trans. Circuits & Systems I*, 50(7):941–945, July 2003.

[136] J. Hynecek. Analysis of the photosite reset in FGA image sensors. *IEEE Trans. Electron Dev.*, 37(10):2193–2200, October 1990.

[137] M. Waeny, S. Tanner, S. Lauxtermann, N. Blanc, M. Willemin, M. Rechsteiner, E. Doering, J. Grupp, P. Seitz, F. Pellandini, and M. Ansorge. High sensitivity and high dynamic, digital CMOS imager. In *Proc. SPIE*, volume 406, pages 78–84, May 2001.

[138] S. Mendis, S.E. Kemeny, and E.R. Fossum. CMOS active pixel image sensor. *IEEE Trans. Electron Dev.*, 41(3):452–453, March 1994.

[139] R.H. Nixon, S.E. Kemeny, B. Pain, C.O. Staller, and E.R. Fossum. 256×256 CMOS active pixel sensor camera-on-a-chip. *IEEE J. Solid-State Circuits*, 31(12):2046–2050, December 1996.

[140] T. Sugiki, S. Ohsawa, H. Miura, M. Sasaki, N. Nakamura, I. Inoue, M. Hoshino, Y. Tomizawa, and T. Arakawa. A 60 mW 10 b CMOS image sensor with column-to-column FPN reduction. In *Dig. Tech. Papers Int'l Solid-State Circuits Conf. (ISSCC)*, pages 108–109, February 2000.

[141] M.F. Snoeij, A.J.P. Theuwissen, K.A.A. Makinwa, and J.H. Huijsing. A CMOS Imager With Column-Level ADC Using Dynamic Column Fixed-Pattern Noise Reduction. *IEEE J. Solid-State Circuits*, 41(12):3007–3015, December 2006.

[142] M. Furuta, S. Kawahito, T. Inoue, and Y. Nishikawa. A cyclic A/D converter with pixel noise and column-wise offset canceling for CMOS image sensors. In *Proc. European Solid-State Circuits Conf. (ESSCIRC)*, pages 411–414, Grenoble, France, September 2005.

[143] S. Smith, J. Hurwitz, M. Torrie, D. Baxter, A. Holmes, M. Panaghiston, R. Henderson, A. Murray, S. Anderson, and P. Denyer. A single-chip 306 × 244-pixel CMOS NTSC video camera. In *Dig. Tech. Papers Int'l Solid-State Circuits Conf. (ISSCC)*, pages 170–171, February 1998.

[144] M.J. Loinaz, K.J. Singh, A.J. Blanksby, D.A. Inglis, K. Azadet, and B.D. Ackland. A 200-mW, 3.3-V, CMOS color camera IC producing 352×288 24-b video at 30 frames/s. *IEEE J. Solid-State Circuits*, 33(12):2092–2103, December 1998.

[145] S. Matsuo, T.J. Bales, M. Shoda, S. Osawa, K. Kawamura, A. Andersson, M. Haque, H. Honda, B. Almond, Y. Mo, J. Gleason, T. Chow, and I. Takayanagi. 8.9-Megapixel Video Image Senso With 14-b Column-Parallel SA-ADC. *IEEE Trans. Electron Dev.*, 56(11):2380–2389, November 2009.

[146] H. Honda, S. Osawa, M. Shoda, E. Pages, T. Sato, N. Karasawa, B. Leichner, J. Schoper, E.S. Gattuso, D. Pates, J. Brooks, S. Johnson, and I. Takayanagi. A 1-inch Optical Format, 14.2M-pixel, 80fps CMOS Image Sensor with a Pipelined Pixel Reset and Readout Operation. In *Dig. Tech. Papers Symp. VLSI Circuits*, pages 2–1, August 2013.

[147] S. Yoshihara, M. Kikuchi, Y. Ito, Y. Inada, S. Kuramochi, H. Wakabayashi, M. Okano, K. Koseki, H. Kuriyama, J. Inutsuka, A. Tajima, T. Nakajima, Y. Kudoh, F. Koga, Y. Kasagi, S. Watanabe, and T. Nomoto. A 1/1.8-inch 6.4MPixel 60 frames/s CMOS Image Sensor with Seamless Mode Change. In *Dig. Tech. Papers Int'l Solid-State Circuits Conf. (ISSCC)*, page 27.1, February 2006.

[148] T. Toyama, K. Mishina, H. Tsuchiya, T. Ichikawa, H. Iwaki, Y. Gendai, H. Murakami, K. Takamiya, H. Shiroshita, Y. Muramatsu, and T. Furusawa. A 17.7Mpixel 120fps CMOS Image Sensor with 34.8Gb/s Readout. In *Dig. Tech. Papers Int'l Solid-State Circuits Conf. (ISSCC)*, page 23.11, February 2011.

[149] S. Decker, D. McGrath, K. Brehmer, and C.G. Sodini. A 256×256 CMOS imaging array with wide dynamic range pixels and column-parallel digital output. *IEEE J. Solid-State Circuits*, 33(12):2081 – 2091, December 1998.

[150] T. Watabe, K. Kitamura, T. Sawamoto, T. Kosugi, T. Akahori, T. Iida, K. Isobe, T. Watanabe, H. Shimamoto, H. Ohtake, S. Aoyama, S. Kawahito, and N. Egami. A 33Mpixel 120fps CMOS Image Sensor Using 12b Column-Parallel Pipelined Cyclic ADCs. In *Dig. Tech. Papers Int'l Solid-State Circuits Conf. (ISSCC)*, page 23.5, February 2012.

[151] Y. Chae, J. Cheon, S. Lim, M. Kwon, K. Yoo, W. Jung and D.-H. Lee, S. Ham, and G. Han. A 2.1 M Pixels, 120 Frame/s CMOS Image Sensor With Column-Parallel $\Delta\Sigma$ ADC Architecture. *IEEE J. Solid-State Circuits*, 46(1):236 – 247, January 2011.

[152] J. Kim, W. Jung, S. Lim, Y. Park, W. Choi, Y. Kim, C. Kang, J. Shin, K. Choo, W. Lee, J. Heo, B. Kim, S. Kim, M. Kwon, K. Yoo, J. Seo, S.n Ham, C. Choi, and G. Han. A 14b Extended Counting ADC Implemented in a 24MPixel APS-C CMOS Image Sensor. In *Dig. Tech. Papers Int'l Solid-State Circuits Conf. (ISSCC)*, page 22.6, February 2012.

[153] S. Kawahito. Column-Parallel ADCs for CMOS Image Sensors and Their FoM-Based Evaluations. *IEICE Trans. Electron.*, E101-C(7):444 – 456, July 2018.

[154] T. Arai, T. Yasue, K. Kitamura, H. Shimamoto, T. Kosugi, S.-W. Jun, S. Aoyama, M.-C. Hsu, Y. Yamashita, H. Sumi, and S. Kawahito. A 1.1-μm 33-Mpixel 240-fps 3-D-Stacked CMOS Image Sensor With Three-Stage Cyclic-Cyclic-SAR Analog-to-Digital Converters. *IEEE Trans. Electron Dev.*, 64(12):4992–5000, December 2017.

[155] D. Yang, B. Fowler, , and A. El Gamal. A Nyquist Rate Pixel Level ADC for CMOS Image Sensors. *IEEE J. Solid-State Circuits*, 34(3):348–356, March 1999.

[156] F. Andoh, H. Shimamoto, and Y. Fujita. A Digital Pixel Image Sensor for Real-Time Readout. *IEEE Trans. Electron Dev.*, 47(11):2123–2127, November 2000.

[157] M. Willemin, N. Blanc, G.K. Lang, S. Lauxtermann, P. Schwider, P. Seitz, and M. Wäny. Optical characterization methods for solid-state image sensors. *Optics and Lasers Eng.*, 36(2):185–194, 2001.

[158] G.R. Hopkinson, T.M. Goodman, and S.R. Prince. *A guide to the use and calibration of detector array equipment.* SPIE Press, Bellingham, Washington, 2004.

[159] M.F. Snoeij, A. Theuwissen, K. Makinwa, and J.H. Huijsing. A CMOS Imager with Column-Level ADC Using Dynamic Column FPN Reduction. In *Dig. Tech. Papers Int'l Solid-State Circuits Conf. (ISSCC)*, pages 2014–2023, February 2006.

[160] J.E. Carnes and W.F. Kosonocky. Noise source in charge-coupled devices. *RCA Review*, 33(2):327–343, June 1972.

[161] Y.P. Tsividis. *Operation and Modeling of the MOS Transistor.* McGraw-Hill, New York, USA, 1987.

[162] Noise Sources in Bulk CMOS by K.H. Lundberg. http://web.mit.edu/klund/www/papers/UNP_noise.pdf.

[163] M. Kasano, Y. Inaba, M. Mori, S. Kasuga, T. Murata, and T. Yamaguchi. A 2.0-μm Pixel Pitch MOS Image Sensor With 1.5 Transistor/Pixel and an Amorphous Si Color Filter. *IEEE Trans. Electron Dev.*, 53(4):611–617, April 2006.

[164] B.E. Bayer. Color imaging array. US patent 3,971,065, July 1976.

[165] B.K. Gunturk, J. Glotzbach, Y. Altunbasak, R.W. Schafer, and R.M. Mersereau. Demosaicking: color filter array interpolation. *IEEE Signal Processing Magazine*, 22(1):44–54, January 2005.

[166] Yoshitaka Egawa ; Nagataka Tanaka ; Nobuhiro Kawai ; Hiromichi Seki ; Akira Nakao ; Hiroto Honda ; Yoshinori Iida ; Makoto Monoi. Operation and performance of a color image sensor with layered photodiodes. In *Proc. SPIE*, volume 5074 of *SPIE AeroSense*, pages 5210–14 – 27, Orlando, FL, April 2003.

[167] S. Koyama, Y. Inaba, M. Kasano, and T. Murata. A Day and Night Vision MOS Imager With Robust Photonic-Crystal-Based RGB-and-IR. *IEEE Trans. Electron Dev.*, 55(5):754–759, March 2008.

[168] S. Yoshimura, T. Sugiyama, K. Yonemoto, and K. Ueda. A 48 kframe/s CMOS image sensor for real-time 3-D sensing and motion detection. In *Dig. Tech. Papers Int'l Solid-State Circuits Conf. (ISSCC)*, pages 94 – 95, February 2001.

[169] J. Nakamura, B. Pain, T. Nomoto, T. Nakamura, and Eric R. Fossum. On-Focal-Plane Signal Processing for Current-Mode Active Pixel Sensors. *IEEE Trans. Electron Dev.*, 1997.

[170] Y. Huanga and R.I. Hornsey. Current-mode CMOS image sensor using lateral bipolar phototransistors. *IEEE Trans. Electron Dev.*, 50(12):2570 – 2573, December 2003.

[171] M.A. Szelezniak, G.W. Deptuch, F. Guilloux, S. Heini, and A. Himmi. Current Mode Monolithic Active Pixel Sensor With Correlated Double Sampling for Charged Particle Detection. *IEEE Sensors Journal*, 7(1):137–150, January 2007.

[172] F. Boussaid, A. Bermak, and A. Bouzerdoum. An Ultra-Low power operating technique for Mega-pixels current-mediated CMOS imagers. *IEEE Trans. Consumer Electron.*, 50(1):46–53, February 2004.

[173] A. Tang, C. Yuan, and A. Bermak. An Ultra-low Power Current-Mode CMOS Image Sensor with Energy Harvesting Capability. In *Proc. European Solid-State Circuits Conf. (ESSCIRC)*, September 2010.

[174] M. Loose, K. Meier, and J. Schemmel. A self-calibrating single-chip CMOS camera with logarithmic response. *IEEE J. Solid-State Circuits*, 36(4):586–596, April 2001.

[175] Y. Oike, M. Ikeda, and K. Asada. High-Sensitivity and Wide-Dynamic-Range Position Sensor Using Logarithmic-Response and Correlation Circuit. *IEICE Trans. Electron.*, E85-C(8):1651–1658, August 2002.

[176] L.-W. Lai, C.-H. Lai, and Y.-C. King. A Novel Logarithmic Response CMOS Image Sensor With High Output Voltage Swing and In-Pixel Fixed-Pattern Noise Reduction. *IEEE Sensors Journal*, 4(1):122–126, February 2004.

[177] B. Choubey, S. Aoyoma, S. Otim, D. Joseph, and S. Collins. An Electronic-Calibration Scheme for Logarithmic CMOS Pixels. *IEEE Sensors Journal*, 6(4):950–956, August 2006.

[178] S. Chamberlain and J.P.Y. Lee. A novel wide dynamic range silicon photodetector and linear imaging array. *IEEE J. Solid-State Circuits*, 19(1), February 1984.

[179] T. Kakumoto, S. Yano, M. Kusudaa, K. Kamon, and Y. Tanaka. Logarithmic conversion CMOS image sensor with FPN cancellation and integration circuits. *J. Inst. Image Information & Television Eng.*, 57(8):1013–1018, August 2003.

[180] B. Fowler, J. Balicki, D. How, and M. Godfrey. Low FPN High Gain Capacitive Transimpedance Amplifier for Low Noise CMOS Image Sensors. In *Proc. SPIE*, volume 4306, pages 68–77, San Francisco, CA, February 2001.

[181] K. Murari, R. Etienne-Cummings, N.V. Thakor, and G. Cauwenberghs. A CMOS In-Pixel CTIA High-Sensitivity Fluorescence Imager. *IEEE Trans. Biomedical Circuits & Sytems*, 5(5):449–458, October 2011.

[182] Ruoyu Xu, Wai Chiu Ng, Jie Yuan, Shouyi Yin, and Shaojun Wei. A 1/2.5 inch VGA 400 fps CMOS Image Sensor With High Sensitivity for Machine Vision. *IEEE J. Solid-State Circuits*, 49(10):2342–2351, October 2014.

[183] X. Qian, H. Yu, and S. Chen. A Global-Shutter Centroiding Measurement CMOS Image Sensor With Star Region SNR Improvement for Star Trackers. *IEEE Trans. Circuits & Systems Video Tech.*, 26(8):1555–1562, August 2016.

[184] D. Stoppa, N. Massari, L. Pancheri, M. Malfatti, M. Perenzoni, and L. Gonzo. A Range Image Sensor Based on 10-μm Lock-In Pixels in 0.18-μm CMOS Imaging Technology. *IEEE J. Solid-State Circuits*, 46(1):248–258, January 2011.

[185] B. Buxbaum, R. Schwarte, and T. Ringbeck. PMD-PLL: receiver structure for incoherent communication and ranging systems. In *Proc. SPIE*, volume 3850, 1999.

[186] W. Van Der Tempel, D. Van Nieuwenhove, R. Grootjans, and M. Kuijk. A Range Image Sensor Based on 10-μm Lock-In Pixels in 0.18-μm CMOS Imaging Technology. *Jpn. J. Appl. Phys.*, 46(4B):2377–2380, April 2007.

[187] W. van der Tempel, A. Ercan, T. Finateu, K. Fotopoulou, C. Mourad, F. Agavriloaie, S. Resimont, L. Cutrignelli, P. Thury, C.E. Medina, S. Xiao, J.-L. Loheac, J. Perhavc, T. Van de Hauwe, V. Belokonskiy, L. Bossuyt, W. Aerts, M. Pauwels, and D. Van Nieuwenhove. A 320x240 10um CAPD ToF image sensor with improved performance. *Int'l Image Sensor Workshop*, pages 137–2380, April 2017.

[188] S. Kawahito, G. Baek, Z. Li, S.-M. Han, M.-W. Seo, K. Yasutomi, and K. Kagawa. CMOS Lock-in Pixel Image Sensors with Lateral Electric Field Control for Time-Resolved Imaging. In *Int'l Image Sensor Workshop*, Snowbird, Utah USA, June 2013.

[189] X. Liu and A. El Gamal. Photocurrent Estimation for a Self-Reset CMOS Image Sensor. In *Proc. SPIE*, volume 4669, pages 304–312, San Jose, CA, 2002.

[190] K.P. Frohmader. A novel MOS compatible light intensity-to-frequency converter suited for monolithic integration. *IEEE J. Solid-State Circuits*, 17(3):588–591, June 1982.

[191] R. Müller. I^2/L timing circuit for the 1 ms-10 s range. *IEEE J. Solid-State Circuits*, 12(2):139–143, April 1977.

[192] V. Brajovic and T. Kanade. A sorting image sensor: An example of massively parallel intensity-to-time processing for low-latency computational

sensors. In *Proc. IEEE Int'l Conf. Robotics & Automation*, pages 1638–1643, Minneapolis, MN, April 1996.

[193] M. Nagata, J. Funakoshi, and A. Iwata. A PWM Signal Processing Core Circuit Based on a Switched Current Integration Technique. *IEEE J. Solid-State Circuits*, 1998.

[194] M. Nagata, M. Homma, N. Takeda, T. Morie, and A. Iwata. A smart CMOS imager with pixel level PWM signal processing. In *Dig. Tech. Papers Symp. VLSI Circuits*, pages 141 – 144, June 1999.

[195] M. Shouho, K. Hashiguchi, K. Kagawa, and J. Ohta. A Low-Voltage Pulse-Width-Modulation Image Sensor. In *IEEE Workshop on Charge-Coupled Devices & Advanced Image Sensors*, pages 226–229, Karuizawa, Japan, June 2005.

[196] S. Shishido, I. Nagahata, T. Sasaki, K. Kagawa, M. Nunoshita, and J. Ohta. Demonstration of a low-voltage three-transistor-per-pixel CMOS imager based on a pulse-width-modulation readout scheme employed with a one-transistor in-pixel comparator. In *Proc. SPIE*, San Jose, CA, 2007. Electronic Imaging.

[197] S. Kleinfelder, S.-H. Lim, X. Liu, and A. El Gamal. A 10 000 Frames/s CMOS Digital Pixel Sensor. *IEEE J. Solid-State Circuits*, 36(12):2049–2059, December 2001.

[198] W. Biderman, A. El Gamal, S. Ewedemi, J. Reyneri, H. Tian, D. Wile, and D. Yang. A 0.18 μm High Dynamic Range NTSC/PAL Imaging System-on-Chip with Embedded DRAM Frame Buffer. In *Dig. Tech. Papers Int'l Solid-State Circuits Conf. (ISSCC)*, pages 212–213, 2003.

[199] J.G. Nicholls, A.R. Martin, B.G. Wallace, and P.A. Fuchs. *From Neuro To Brain*. Sinauer Associates, Inc., Sunderland, MA, 4th edition, 2001.

[200] W. Maass and C.M. Bishop, editors. *Pulsed Neural Networks*. The MIT Press, Cambridge, MA, 1999.

[201] T. Lehmann and R. Woodburn. Biologically-inspired learning in pulsed neural networks. In G. Cauwenberghs and M.A. Bayoumi, editors, *Learning on silicon: adaptive VLSI neural systems*, pages 105–130. Kluwer Academic Pub., Norwell, MA, 1999.

[202] K. Kagawa, K. Yasuoka, D. C. Ng, T. Furumiya, T. Tokuda, J. Ohta, and M. Nunoshita. Pulse-domain digital image processing for vision chips employing low-voltage operation in deep-submicron technologies. *IEEE Selected Topic Quantum Electron.*, 10(4):816–828, July 2004.

[203] T. Hammadou. Pixel Level Stochastic Arithmetic For Intelligent Image Capture. In *Proc. SPIE*, volume 5301, pages 161–167, San Jose, CA, January 2004.

[204] W. Yang. A wide-dynamic-range, low-power photosensor array. In *Dig. Tech. Papers Int'l Solid-State Circuits Conf. (ISSCC)*, pages 230–231, February 1994.

[205] K. Kagawa, S. Yamamoto, T. Furumiya, T. Tokuda, M. Nunoshita, and J. Ohta. A pulse-frequency-modulation vision chip using a capacitive feedback reset with an in-pixel 1-bit image processing. In *Proc. SPIE*, volume 6068, pages 60680C–1–60680C–9, San Jose, January 2006.

[206] X. Wang, W. Wong, and R. Hornsey. A High Dynamic Range CMOS Image Sensor With Inpixel Light-to-Frequency Conversion. *IEEE Trans. Electron Dev.*, 53(12):2988 – 2992, December 2006.

[207] E. Culurciello, R. Etienne-Cummings, and K.A. Boahen. A Biomorphic Digital Image Sensor. *IEEE J. Solid-State Circuits*, 38(2):281–294, February 2003.

[208] T. Teixeira, A.G. Andreou, and E. Culurciello. An Address-Event Image Sensor Network. In *Int'l Symp. Circuits & Systems (ISCAS)*, pages 644–647, Kobe, Japan, May 2005.

[209] T. Teixeira, E. Culurciello, and A.G. Andreou. An Address-Event Image Sensor Network. In *Int'l Symp. Circuits & Systems (ISCAS)*, pages 4467–4470, Kos, Greece, May 2006.

[210] M.L. Simpson, G.S. Sayler, G. Patterson, D.E. Nivens, E.K. Bolton, J.M. Rochelle, and J.C. Arnott. An integrated CMOS microluminometer for low-level luminescence sensing in the bioluminescent bioreporter integrated circuit. *Sensors & Actuators B*, 72:134–140, 2001.

[211] E.K. Bolton, G.S. Sayler, D.E. Nivens, J.M. Rochelle, S. Ripp, and M.L. Simpson. Integrated CMOS photodetectors and signal processing for very low level chemical sensing with the bioluminescent bioreporter integrated circuits. *Sensors & Actuators B*, 85:179–185, 2002.

[212] J. Ohta, N. Yoshida, K. Kagawa, and M. Nunoshita. Proposal of Application of Pulsed Vision Chip for Retinal Prosthesis. *Jpn. J. Appl. Phys.*, 41(4B):2322–2325, April 2002.

[213] K. Kagawa, K. Isakari, T. Furumiya, A. Uehara, T. Tokuda, J. Ohta, and M. Nunoshita. Pixel design of a pulsed CMOS image sensor for retinal prosthesis with digital photosensitivity control. *Electron. Lett.*, 39(5):419–421, May 2003.

[214] A. Uehara, K. Kagawa, T. Tokuda, J. Ohta, and M. Nunoshita. Back-illuminated pulse-frequency-modulated photosensor using a silicon-on-sapphire technology developed for use as an epi-retinal prosthesis device. *Electron. Lett.*, 39(15):1102–1104, July 2003.

[215] David C. Ng, K. Isakari, A. Uehara, K. Kagawa, T. Tokuda, J. Ohta, and M. Nunoshita. A study of bending effect on pulsed frequency

modulation based photosensor for retinal prosthesis. *Jpn. J. Appl. Phys.*, 42(12):7621–7624, December 2003.

[216] T. Furumiya, K. Kagawa, A. Uehara, T. Tokuda, J. Ohta, and M. Nunoshita. 32 × 32-pixel pulse-frequency-modulation based image sensor for retinal prosthesis. *J. Inst. Image Information & Television Eng.*, 58(3):352–361, March 2004. In Japanese.

[217] K. Kagawa, N. Yoshida, T. Tokuda, J. Ohta, and M. Nunoshita. Building a Simple Model of A Pulse-Frequency-Modulation Photosensor and Demonstration of a 128 × 128-pixel Pulse-Frequency-Modulation Image Sensor Fabricated in a Standard 0.35-μm Complementary Metal-Oxide Semiconductor Technology. *Opt. Rev.*, 11(3):176–181, May 2004.

[218] A. Uehara, K. Kagawa, T. Tokuda, J. Ohta, and M. Nunoshita. A high-sensitive digital photosensor using MOS interface-trap charge pumping. *IEICE Electronics Express*, 1(18):556–561, December 2004.

[219] T. Furumiya, D. C. Ng, K. Yasuoka, K. Kagawa, T. Tokuda, M. Nunoshita, and J. Ohta. Functional verification of pulse frequency modulation-based image sensor for retinal prosthesis by *in vitro* electrophysiological experiments using frog retina. *Biosensors & Bioelectron.*, 21(7):1059–1068, January 2006.

[220] S. Yamamoto, K. Kagawa, T. Furumiya, T. Tokuda, M. Nunoshita, and J. Ohta. Prototyping and evaluation of a 32 × 32-pixel pulse-frequency-modulation vision chip with capacitive-feedback reset. *J. Inst. Image Information & Television Eng.*, 60(4):621–626, April 2006. In Japanese.

[221] T. Furumiya, S. Yamamoto, K. Kagawa, T. Tokuda, M. Nunoshita, and J. Ohta. Optimization of electrical stimulus pulse parameter for low-power operation of a retinal prosthetic device. *Jpn. J. Appl. Phys.*, 45(19):L505 – L507, May 2006.

[222] D. C. Ng, T. Furumiya, K. Yasuoka, A. Uehara, K. Kagawa, T. Tokuda, M. Nunoshita, and J. Ohta. Pulse Frequency Modulation-based CMOS Image Sensor for Subretinal Stimulation. *IEEE Trans. Circuits & Systems II*, 53(6):487–491, June 2006.

[223] J. Ohta, T. Tokuda, K. Kagawa, T. Furumiya, A. Uehara, Y. Terasawa, M. Ozawa, T. Fujikado, and Y. Tano. Silicon LSI-Based Smart Stimulators for Retinal Prosthesis. *IEEE Eng. Medicine & Biology Magazine*, 25(5):47–59, October 2006.

[224] J. Deguchi, T. Watanabe, T. Nakamura, Y. Nakagawa, T. Fukushima, S. Jeoung-Chill, H. Kurino, T. Abe, M. Tamai, and M. Koyanagi. Three-Dimensionally Stacked Analog Retinal Prosthesis Chip. *Jpn. J. Appl. Phys.*, 43(4B):1685–1689, April 2004.

[225] D. Ziegler, P. Linderholm, M. Mazza, S. Ferazzutti, D. Bertrand, A.M. Ionescu, and Ph. Renaud. An active microphotodiode array of oscillating pixels for retinal stimulation. *Sensors & Actuators A*, 110:11–17, 2004.

[226] M. Mazza, P. Renaud, D.C. Bertrand, and A.M. Ionescu. CMOS Pixels for Subretinal Implantable Prothesis. *IEEE Sensors Journal*, 5(1):32–27, February 2005.

[227] M.L. Prydderch, M.J. French, K. Mathieson, C. Adams, D. Gunning, J. Laudanski, J.D. Morrison, A.R. Moodie, and J. Sinclair. A CMOS Active Pixel Sensor for Retinal Stimulation. In *Proc. SPIE*, pages 606803–1 – 606803–9, San Jose, 2006. Electronic Imaging.

[228] K. Sasagawa, T. Yamaguchi, M. Haruta, Y. Sunaga, Hironari Takeahra, Hiroaki Takehara, T. Noda, T. Tokuda, and J. Ohta. An Implantable CMOS Image Sensor With Self-Reset Pixels for Functional Brain Imaging. *IEEE Trans. Electron Dev.*, 63(1):215–222, January 2016.

[229] T. Yamaguchi, Hiroaki Takeahra, Y. Sunaga, M. Haruta, M. Mototyama, Y. Ohta, T. Noda K. Sasagawa, T. Tokuda, and J. Ohta. Implantable self-reset CMOS image sensor and its application to hemodynamic response detection in living mouse brain. *Jpn. J. Appl. Phys.*, 55:04EM02, April 2016.

[230] A. Bermak, A. Bouzerdou, and K. Eshraghian. A vision sensor with on-pixel ADC and in-built light adaptation mechanism. *Microelectronics J.*, 33(2):1091–1096, December 2002.

[231] J. Yuan, H.Y. Chan, S.W. Fung, and B. Liu. An activity-triggered 95.3 dB DR-75.6 dB THD CMOS imaging sensor with digital calibration. *IEEE J. Solid-State Circuits*, 44(10):2834–2843, October 2009.

[232] D. Park, J. Rhee, and Y. Joo. Wide dynamic-range CMOS image sensor using self-reset technique. *IEEE Electron Device Lett.*, 28(10):890–892, October 2007.

[233] S. Koppa, D. Park, Y. Joo, and S. Jung. A 105.6 dB DR and 65 dB peak SNR self-reset CMOS image sensor using a Schmitt trigger circuit. In *IEEE Int'l Midwest Symp. Cir. Sys. (MWSCAS)*, Seoul, South Korea, August 2011.

[234] D. Bronzi, F. Villa, S. Tisa, A. Tosi, and F. Zappa. SPAD Figures of Merit for Photon-Counting, Photon-Timing, and Imaging Applications: A Review. *IEEE Sensors Journal*, 16(1):3–12, January 2016.

[235] A.C. Ulku, C. Bruschini, I.M. Antolovic, Y. Kuo, R. Ankri, S. Weiss, X. Michalet, and E. Charbon. A 512 × 512 SPAD Image Sensor with Integrated Gating for Widefield FLIM. *IEEE Selected Topic Quantum Electron.*, 2018.

[236] C. Shi, J. Yang, Y. Han, Z. Cao, Q. Qin, L. Liu, N.-J. Wu, and Z. Wang. A 1000 fps Vision Chip Based on a Dynamically Reconfigurable Hybrid

Architecture Comprising a PE Array Processor and Self-Organizing Map Neural Network. *IEEE J. Solid-State Circuits*, 49(9):2067–2082, September 2014.

[237] S. Kagami, T. Komuro, and M. Ishikawa. A Software-Controlled Pixel-Level ADC Conversion Method for Digital Vision Chips. In *IEEE Workshop on Charge-Coupled Devices & Advanced Image Sensors*, Elmau, Germany, May 2003.

[238] J.B. Kuoa and S.-C. Lin. *Low-voltage SOI CMOS VLSI devices and circuits*. John Wiley & Sons, Inc., New York, NY, 2001.

[239] A. Afzalian and D. Flandre. Modeling of the bulk versus SOI CMOS performances for the optimal design of APS circuits in low-power low-voltage applications. *IEEE Trans. Electron Dev.*, 2003.

[240] K. Kioi, T. Shinozaki, S. Toyoyama, K. Shirakawa, K. Ohtake, and S. Tsuchimoto. Design and implementation of a 3D-LSI image sensing processor. *IEEE J. Solid-State Circuits*, 27(8):1130 – 1140, August 1992.

[241] V. Suntharalingam, R. Berger, J.A. Burns, C.K. Chen, C.L. Keast, J.M. Knecht, R.D. Lambert, K.L. Newcomb, D.M. O'Mara, D.D. Rathman, D.C. Shaver, A.M. Soares, C.N. Stevenson, B.M. Tyrrell, K. Warner, B.D. Wheeler, D.-R.W. Yost, and D.J. Young. Megapixel CMOS image sensor fabricated in three-dimensional integrated circuit technology. In *Dig. Tech. Papers Int'l Solid-State Circuits Conf. (ISSCC)*, pages 356 – 357, February 2005.

[242] M. Goto, K. Hagiwara, Y. Iguchi, H. Ohtake, T. Saraya, M. Kobayashi, E. Higurashi, H. Toshiyoshi, and T. Hiramoto. Pixel-Parallel 3-D Integrated CMOS Image Sensors With Pulse Frequency Modulation A/D Converters Developed by Direct Bonding of SOI Layers. *IEEE Trans. Electron Dev.*, 62(11):3350–3355, November 2015.

[243] S. Iwabuchi, Y. Maruyama, Y. Ohgishi, M. Muramatsu, N. Karasawa, and T. Hirayama. A Back-Illuminated High-Sensitivity Small-Pixel Color CMOS Image Sensor with Flexible Layout of Metal Wiring. In *Dig. Tech. Papers Int'l Solid-State Circuits Conf. (ISSCC)*, pages 1171–1178, February 2006.

[244] C. Xu, W. Zhang, and M. Chan. A low voltage hybrid bulk/SOI CMOS active pixel image sensor. *IEEE Electron Device Lett.*, 22(5):248 – 250, May 2001.

[245] C. Xu, C. Shen, W. Wu, and M. Chan. Backside-Illuminated Lateral PIN Photodiode for CMOS Image Sensor on SOS Substrate. *IEEE Trans. Electron Dev.*, 52(6):1110–1115, June 2005.

[246] Y. Arai, T. Miyoshi, Y. Unno, T. Tsuboyama, S. Terada, Y. Ikegami, R. Ichimiya, T. Kohriki, K. Tauchi, Y. Ikemoto, Y. Fujita, T. Uchida, K. Hara, H. Miyake, M. Kochiyama, T. Sega, K. Hanagaki, M. Hirose, J. Uchida, Y. Onuki, Y. Horii, H. Yamamoto, T. Tsuru, H. Matsumoto, S. G. Ryu, R. Takashima, A. Takeda, H. Ikeda, D. Kobayashi, T. Wada, and H. Nagata.

Development of SOI pixel process technology. *Nuclear Instruments and Methods in Physics Research A*, 636:S31–S36, 2011.

[247] T. Ishikawa, M. Ueno, Y. Nakaki, K. Endo, Y. Ohta, J. Nakanishi, Y. Kosasayama, H. Yagi, T. Sone, and M. Kimata. Performance of 320 × 240 Uncooled IRFPA with SOI Diode Detectors. In *Proc. SPIE*, volume 4130, pages 152–159, 2000.

[248] Y. Onuki, H. Katsurayama, Y. Ono, H. Yamamoto, Y. Arai, Y. Fujita, R. Ichimiya, Y. Ikegami, Y. Ikemoto, T. Kohriki, T. Miyoshi, K. Tauchi, S. Terada, T. Tsuboyama, Y. Unno, T. Uchida, K. Hara, K. Shinsho, A. Takeda, K. Hanagaki, T. G. Tsuru, S.G. Ryu, S. Nakashima, H. Matsumoto, R. Takashima, H. Ikeda, D. Kobayashi, T. Wada, T. Hatsui, T. Kudo, A. Taketani, K. Kobayashi, Y. Kirihara, S. Ono, M. Omodani, T. Kameshima, Y. Nagatomo, H. Kasai, N. Kuriyama, N. Miura, and M. Okihara. SOI detector developments. In *Proceedings of Science, PoS (Vertex 2011)*, volume 043, pages 1–8, 2011.

[249] E. Culurciello. *Silicon-on-Sapphire Circuits and Systems – Sensor and Biosensor Interfaces*. McGraw-Hill, Columbus, OH, 2009.

[250] A.G. Andreou, Z.K. Kalayjian, A. Apsel, P.O. Pouliquen, R.A. Athale, G. Simonis, and R. Reedy. Silicon on sapphire CMOS for optoelectronic microsystems. *IEEE Circuits & Systems Magazine*, 2001.

[251] E. Culurciello and A.G. Andreou. 16 × 16 pixel silicon on sapphire CMOS digital pixel photosensor array. *Electron. Lett.*, 40(1):66–68, January 2004.

[252] A. Fish, E. Avner, and O. Yadid-Pecht. Low-power global/rolling shutter image sensors in silicon on sapphire technology. In *Int'l Symp. Circuits & Systems (ISCAS)*, pages 580–583, Kobe, Japan, May 2005.

[253] S. Yokogawa, I. Oshiyama, H. Ikeda, Y. Ebiko, T. Hirano, S. Saito, T. Oinoue, Y. Hagimoto, and H. Iwamoto. IR sensitivity enhancement of CMOS Image Sensor with diffractive light trapping pixels. *Sci. Rep.*, 7:3832, June 2017.

[254] Z. Huang, J.E. Carey, M. Liu, X. Guo, E. Mazur, and J.C. Campbell. Microstructured silicon photodetector. *Appl. Phys. Lett.*, 89:033506, July 2006.

[255] S.D. Gunapala, S.V. Bandara, J.K. Liu, Sir B. Rafol, and J.M. Mumolo. 640 × 512 Pixel Long-Wavelength Infrared Narrowband, Multiband, and Broadband QWIP Focal Plane Arrays. *IEEE Trans. Electron Dev.*, 50(12):2353–2360, December 2003.

[256] M. Kimata. Infrared Focal Plane Arrays. In H. Baltes, W. Gopel, and J. Hesse, editors, *Sensors Update*, volume 4, pages 53–79. Wiley-VCH, 1998.

[257] C.-C. Hsieh, C.-Y. Wu, F.-W. Jih, and T.-P. Sun. Focal-Plane-Arrays and CMOS Readout Techniques of Infrared Imaging Systems. *IEEE Trans. Circuits & Systems Video Tech.*, 7(4):594–605, August 1997.

[258] P.E. Malinowski, E. Georgitzikis, J. Maes, I. Vamvaka, F. Frazzica, J. Van Olmen, P. De Moor, P. Heremans, Z. Hens, and D. Cheyns. Thin-Film Quantum Dot Photodiode for Monolithic Infrared Image Sensors. *Sensors*, 17:2867, 2017.

[259] M. Kimata. Silicon infrared focal plane arrays. In M. Henini and M. Razeghi, editors, *Handbook of Infrared Detection Technologies*, pages 352–392. Elsevier Science Ltd., 2002.

[260] E. Kasper and K. Lyutovich, editors. *Properties of Silicon Germanium and SiGe:Carbon*. INSPEC, The Institute of Electrical Engineers, London, UK, 2000.

[261] J. Michel, J. Liu, and L.C. Kimerling. High-performance Ge-on-Si photodetectors. *Nature Photonics*, 4:527–534, August 2010.

[262] B. Ackland, C. Rafferty, C. King, I. Aberg, J. O'Neill, T. Sriram, A. Lattes, C. Godek, and S. Pappas. A Monolithic Ge-on-Si CMOS Imager for Short Wave Infrared. In *Int'l Image Sensor Workshop*, Bergen, Norway, June 2009.

[263] N. Na, S.-L. Cheng, H.-D. Liu, M.-J. Yang, C.-Y. Chen, H.-W. Chen, Y.-T. Chou, C.-T. Lin, W.-H. Liu, C.-F. Liang, C.-L. Chen, S.-W. Chu, B.-J. Chen, Y.-F. Lyu, , and S.-L. Chen. High-Performance Germanium-on-Silicon Lock-in Pixels for Indirect Time-of-Flight Applications. In *Tech. Dig. Int'l Electron Devices Meeting (IEDM)*, 2018.

[264] T. Tokuda, D. Mori, K. Kagawa, M. Nunoshita, and J. Ohta. A CMOS image sensor with eye-safe detection function using backside carrier injection. *J. Inst. Image Information & Television Eng.*, 60(3):366–372, March 2006. In Japanese.

[265] T. Tokuda, Y. Sakano, K. Kagawa, J. Ohta, and M. Nunoshita. Backside-hybrid photodetector for trans-chip detection of NIR light. In *IEEE Workshop on Charge-Coupled Devices & Advanced Image Sensors*, Elmau, Germany, May 2003.

[266] S. Iwabuchi, Y. Maruyama, Y. Ohgishi, M. Muramatsu, N. Karasawa, and T. Hirayama. A Back-Illuminated High-Sensitivity Small-Pixel Color CMOS Image Sensor with Flexible Layout of Metal Wiring. In *Dig. Tech. Papers Int'l Solid-State Circuits Conf. (ISSCC)*, San Francisco, CA, USA, February 2006.

[267] H. Wakabayashi, K. Yamaguchi, M. Okano, S. Kuramochi, O. Kumagai, S. Sakane, M. Ito, M. Hatano, M. Kikuchi, Y. Yamagata, T. Shikanai, K. Koseki, K. Mabuchi, Y. Maruyama, K. Akiyama, E. Miyata, T. Honda, M. Ohashi, and T. Nomoto. A 1/2.3-inch 10.3Mpixel 50frame/s Back-Illuminated CMOS Image Sensor. In *Dig. Tech. Papers Int'l Solid-State Circuits Conf. (ISSCC)*, San Francisco, CA, USA, February 2010.

[268] S. Sukegawa, T. Umebayashi, T. Nakajima, H. Kawanobe, K. Koseki, I. Hirota, T. Haruta, M. Kasai, K. Fukumoto, T. Wakano, K. Inoue, H. Takahashi, T. Nagano, Y. Nitta, T. Hirayama, and N. Fukushima. A 1/4-inch 8Mpixel back-illuminated stacked CMOS image sensor. In *Dig. Tech. Papers Int'l Solid-State Circuits Conf. (ISSCC)*, pages 484–486, San Francisco, CA, USA, February 2013.

[269] H. Rhodes, D. Tai, Y. Qian, D. Mao, V. Venezia, W. Zheng, Z. Xiong, C.Y. Liu, K.C. Ku, S. Manabe, A. Shah, S. Sasidhar, P. Cizdziel, Z. Lin, A. Ercan, M. Bikumandla, R. Yang, P. Matagne, C. Yang, H. Yang, T.J. Dai, J. Li, S.G. Wuua, D.N. Yaunga, C.C. Wanga, J.C. Liua, C.S. Tsaia, Y.L.Tua, and T.H. Hsua. The Mass Production of BSI CMOS Image Sensors. In *Int'l Image Sensor Workshop*, Snowbird, Utah USA, June 2013.

[270] S. Choi, S. Lim, M. Lim, H. Bae, K. Choo, J. Kang, S. Lee, S. Kim, J. Moon, K. Son, E. Shim, H. Cho, Y. Kim, S. Ham, J. Ahn, C. Moon, and D. Lee. Back-side illuminated 28M pixel APS C sensor with high performance. In *Int'l Image Sensor Workshop*, Vaals, The Netherlands, June 2015.

[271] T. Haruta, T. Nakajima, J. Hashizume, T. Umebayashi, H. Takahashi, K. Taniguchi, M. Kuroda, H. Sumihiro, K. Enoki, T. Yamasaki, K. Ikezawa, A. Kitahara, M. Zen, M. Oyama, H. Koga, H. Tsugawa, T. Ogita, T. Nagano, S. Takano, and T. Nomoto. A 1/2.3inch 20Mpixel 3-Layer Stacked CMOS Image Sensor with DRAM. In *Dig. Tech. Papers Int'l Solid-State Circuits Conf. (ISSCC)*, pages 76–78, February 2017.

[272] H. Tsugawa, H. Takahashi, R. Nakamura, T. Umebayashi, T. Ogita, H. Okano, K. Iwase, H. Kawashima, T. Yamasaki, D. Yoneyama, J. Hashizume, T. Nakajima, K. Murata, Y. Kanaishi, K. Ikeda, K. Tatani, T. Nagano, H. Nakayama, T. Haruta, and T. Nomoto. Pixel/DRAM/logic 3-layer stacked CMOS image sensor technology. In *Tech. Dig. Int'l Electron Devices Meeting (IEDM)*, pages 3.2.1–3.2.4, December 2017.

[273] H. Sugo, S. Wakashima, R. Kuroda, Y. Yamashita, H. Sumi, T.-J. Wang, P.-S. Chou, M.-C. Hsu, and S. Sugawa. A Dead-time Free Global Shutter CMOS Image Sensor with in-pixel LOFIC and ADC using Pixel-wise Connections. In *Dig. Tech. Papers Symp. VLSI Technology*, pages 1–2, June 2016.

[274] T. Arai, T. Yasue, K. Kitamura, H. Shimamoto, T. Kosugi, S.-W. Jun, S. Aoyama, M.-C. Hsu, Y. Yamashita, H. Sumi, and S. Kawahito. A 1.1-μm 33-Mpixel 240-fps 3-D-Stacked CMOS Image Sensor With Three-Stage Cyclic-Cyclic-SAR Analog-to-Digital Converters. *IEEE Trans. Electron Dev.*, 2017.

[275] T. Takahashi, Y. Kaji, Y. Tsukuda, S. Futami, K. Hanzawa, T. Yamauchi, P.W. Wong, F.T. Brady, P. Holden, T. Ayers, K. Mizuta, S. Ohki, K. Tatani, H. Wakabayashi, and Y. Nitta. A Stacked CMOS Image Sensor With Array-Parallel ADC Architecture. *IEEE J. Solid-State Circuits*, 53(4):1061–1070, April 2018.

[276] H. Seo, S. Aihara, T. Watabe, H. Ohtake, T. Sakai, M. Kubota, N. Egami, T. Hiramatsu, T. Matsuda, M. Furuta, and T. Hirao. A 128 x 96 Pixel Stack-Type Color Image Sensor with B-, G-, R-sensitive organic photoconductive films. In *Int'l Image Sensor Workshop*, 2011.

[277] M. Mori, Y. Hirose, M. Segawa, I. Miyanaga, R. Miyagawa, T. Ueda, H. Nara, H. Masuda, S. Kishimura, T. Sasaki, Y. Kato, Y. Imada, H. Asano, H. Inomata, H. Koguchi, M. Ihama, and Y. Mishima. Thin Organic Photoconductive Film Image Sensors with Extremely High Saturation of 8500 electrons/μm^2. In *Dig. Tech. Papers Symp. VLSI Circuits*, pages T22–T23, 2013.

[278] S. Shishido, Y. Miyake, Y. Sato, T. Tamaki, N. Shimasaki, Y. Sato, M. Murakami, and Y. Inoue. 210ke$^-$ Saturation Signal 3 μ m-Pixel Variable-Sensitivity Global-Shutter Organic Photoconductive Image Sensor for Motion Capture. In *Dig. Tech. Papers Int'l Solid-State Circuits Conf. (ISSCC)*, pages 112–114, February 2016.

[279] K.M. Findlater, D.Renshaw, J.E.D. Hurwitz, R.K. Henderson, T.E.R. Biley, S.G. Smith, M.D. Purcell, and J.M. Raynor. A CMOS Image Sensor Employing a Double Junction Photodiode. In *IEEE Workshop on Charge-Coupled Devices & Advanced Image Sensors*, pages 60–63, Lake Tahoe, NV, June 2001.

[280] D.L. Gilblom, S.K. Yoo, and P. Ventura. Operation and performance of a color image sensor with layered photodiodes. In *Proc. SPIE*, volume 5074 of *SPIE AeroSense*, pages 5210-14 – 27, Orlando, FL, April 2003.

[281] R.B. Merrill. Color Separation in an Active Pixel Cell Imaging Array Using a Triple-Well Structure. US Patent 5,965,875, 1999.

[282] T. Lulé, B. Schneider, and M. Böhm. Design and Fabrication of a High-Dynamic-Range Image Sensor in TFA Technology. *IEEE J. Solid-State Circuits*, 34(5):704–711, May 1999.

[283] M. Sommer, P. Rieve, M. Verhoeven, M. Böhm, B. Schneider, B. van Uffel, and F. Librecht. First Multispectral Diode Color Imager With Three Color Recognition And Color Memory In Each Pixel. In *IEEE Workshop on Charge-Coupled Devices & Advanced Image Sensors*, pages 187–190, Karuizawa, Japan, June 1999.

[284] H. Steibig, R.A. Street, D. Knipp, M. Krause, and J. Ho. Vertically integrated thin-film color sensor arrays for advanced sensing applications. *Appl. Phys. Lett.*, 88:013509, 2006.

[285] Y. Maruyama, K. Sawada, H. Takao, and M. Ishida. The fabrication of filter-less fluorescence detection sensor array using CMOS image sensor technique. *Sensors & Actuators A*, 128:66–70, 2006.

[286] Y. Maruyama, K. Sawada, H. Takao, and M. Ishida. A novel filterless fluorescence detection sensor for DNA analysis. *IEEE Trans. Electron Dev.*, 53(3):553– 558, March 2006.

[287] L.W. Barnes, A. Dereux, and T.W. Ebbesen. Surface plasmon subwavelength optics. *Nature*, 2003.

[288] S. Yokogawa, S.P. Burgos, and H.A. Atwater. Plasmonic Color Filters for CMOS Image Sensor Applications. *Nano Lett.*, 12(8):4349–4354, July 2012.

[289] Q. Chen, D. Das, D. Chitnis, K. Walls, T. D. Drysdale, S. Collins, and D. R. S. Cumming. A CMOS Image Sensor Integrated with Plasmonic Colour Filters. *Plasmonics*, 7:695–699, March 2012.

[290] A. Ono, A. Miyamichi, H. Kamehama, K. Kagawa, K. Yasutomi, and S. Kawahito. Nanostructured metallic color filter for wide-range and multi-band image sensor. In *Int'l Image Sensor Workshop*, pages 59–61, Hiroshima, Japan, May 2017.

[291] A. Miyamichi, A. Ono, H. Kamehama, K. Kagawa, K. Yasutomi, and S. Kawahito. Multi-band plasmonic color filters for visible-to-near-infrared image sensors. *Opt. Express*, 26(19):25178–25187, September 2018.

[292] Y. Inaba, M. Kasano, K. Tanaka, and T. Yamaguchi. Degradation-free MOS image sensor with photonic crystal color filter. *IEEE Electron Device Lett.*, 27(6):457 – 459, June 2006.

[293] H.A. Bethe. Theory of Diffraction by Small Holes. *Phys. Rev.*, 66(7-8):163–182, October 1944.

[294] T. Thio, K.M. Pellerin, R.A. Linke, H.J. Lezec, and T.W. Ebbessen. Enhanced light transmission through a single subwavelength aperture. *Opt. Lett.*, 26, December 2001.

[295] H. Raether. *Surface Plasmons on Smooth and Rough Surfaces and on Gratings*. Springer-Verlag, Berlin, Germany, 1988.

[296] K. Sasagawa, K. Kusawake, K. Kagawa, J. Ohta, and M. Nunoshita. Optical transmission enhancement for an image sensor with a sub-wavelength aperture. In *Int'l Conf. Optics-Photonics Design & Fabrication (ODF)*, pages 163–164, November 2002.

[297] L. Hong, H. Li, H. Yang, and K. Sengupta. Nano-plasmonics and electronics co-integration in CMOS enabling a pill-sized multiplexed fluorescence microarray system. *Nature*, 8(11):5735–5758, November 2018.

[298] S. Nishiwaki, T. Nakamura, M. Hiramoto, T. Fujii, and M. Suzuki. Efficient colour splitters for high-pixel-density image sensors. *Nature Photonics*, 7:240–246, March 2013.

[299] P. Wang and R. Menon. Ultra-high-sensitivity color imaging via a transparent diffractive-filter array and computational optics. *Optica*, 2(11):933–939, November 2015.

[300] Masashi Miyata, Mitsumasa Nakajima, and Toshikazu Hashimoto. High-Sensitivity Color Imaging Using Pixel-Scale Color Splitters Based on Dielectric Metasurfaces. *ACS Photonics*, 6(6):1442–1450, June 2019.

[301] M.F. Land and D.-E. Nilsson. *Animal Eyes*. Oxford University Press, Oxford, UK, 2002.

[302] E. Hecht. *Optics*. Addison-Wesley Pub. Co., Reading, MA, 2nd edition, 1987.

[303] H. Endo. "Phase Detection Pixel Built-in Image Sensor" to Realize High Speed Auto Focus. *J. Inst. Image Information and Television Eng.*, 65(3):290–292, March 2016. Japanese.

[304] M. Kobayashi, M. Johnson, Y. Wada, H. Tsuboi, H. Takada, K. Togo, T. Kishi, H. Takahashi, T. Ichikawa, and S. Inoue. A Low Noise and High Sensitivity Image Sensor with Imaging and Phase-Difference Detection AF in All Pixels. *ITE Trans. Media Tech. Appl.*, 4(2):123–128, February 2016.

[305] S. Yokogawa, I. Hirota, I. Ohdaira, M. Matsumura, A. Morimitsu, H. Takahashi, T. Yamazaki, H. Oyaizu, Y. Incesu, M. Atif, and Y. Nitta. A 4M pixel full-PDAF CMOS image sensor with 1.58 μ m 2X1 On-Chip Micro-Split-Lens technology. In *Int'l Image Sensor Workshop*, Vaals, The Netherlands, June 2015.

[306] K. Aizawa, K. Sakaue, and Y. Suenaga, editors. *Image Processig Technologies, Algorithms, Sensors, and Applications*. Mracel Dekker, Inc., New York, NY, 2004.

[307] K. Yamazawa, Y. Yagi, and M. Yachida. Ominidirectional Imaging with Hyperboloidal Projection. In *Proc. IEEE/RSJ Int'l Conf. Intelligent Robots & Systems*, pages 1029–1034, Yokohama, Japan, July 1993.

[308] J. Ohta, H. Wakasa, K. Kagawa, M. Nunoshita, M. Suga, M. Doi, M. Oshiro, K. Minato, and K. Chihara. A CMOS image sensor for Hyper Omni Vision. *Trans. Inst. Electrical Eng. Jpn., E*, 123-E(11):470–476, November 2003.

[309] H. C. Ko, M. P. Stoykovich, J. Song, V. Malyarchuk, W. M. Choi, C.-J. Yu, J.B Geddes III, J. Xiao, S. Wang, Y. Huang, and J. A. Rogers. A hemispherical electronic eye camera based on compressible silicon optoelectronics. *Nature*, 454:748–753, 2008.

[310] K. Itonaga, T. Arimura, K. Matsumoto, K. Terahata G. Kondo, S. Makimoto, M. Baba, Y .Honda, S. Bori, T. Kai, K. Kasahara, M. Nagano, M. Kimura andY. Kinoshita, E. Kishida, T .Baba, S. Baba, Y. Nomura, N. Tanabe, N. Kimizuka, Y .Matoba, T. Takachi, E. Takagi, T. Haruta, N. Ikebe, K. Matsuda, T. Niimi, T. Ezaki, and T. Hirayama. A Novel Curved CMOS

Image Sensor Integrated with Imaging System. In *Dig. Tech. Papers Symp. VLSI Technology*, June 2014.

[311] B. Guenter, N. Joshi, R. Stoakley, A. Keefe, K. Geary, R. Freeman, J. Hundley, P. Patterson, D. Hammon, G. Herrera, E. Sherman, A. Nowak, R. Schubert, P. Brewer, L. Yang, R. Mott, and G. McKnight. Highly curved image sensors: a practical approach for improved optical performance. *Opt. Express*, 25(12):13010–13023, 2017.

[312] J. Tanidaa, T. Kumagai, K. Yamada, S. Miyatakea, K. Ishida, T. Morimoto, N. Kondou, D. Miyazaki, and Yoshiki Ichioka. Thin observation module by bound optics (TOMBO): concept and experimental verification. *Appl. Opt.*, 40(11):1806–1813, April 2001.

[313] S. Ogata, J. Ishida, and H. Koshi. Optical sensor array in an artificial compound eye. *Opt. Eng.*, 33:3649–3655, November 1994.

[314] J.S. Sanders and C.E. Halford. Design and analysis of apposition compound eye optical sensors. *Opt. Eng.*, 34(1):222–235, January 1995.

[315] K. Hamanaka and H. Koshi. An artificial compound eye using a microlens array and its application to scale-invariant processing. *Opt. Rev.*, 3(4):265–268, 1996.

[316] J. Duparré, P. Dannberg, P. Schreiber, A. Bräuer, and A. Tünnermann. Micro-optically fabricated artificial apposition compound eye. In *Proc. SPIE*, volume 5301, pages 25–33, San Jose, CA, January 2004.

[317] R. Hornsey, P. Thomas, W. Wong, S. Pepic, K. Yip, and R. Kishnasamy. Electronic compound-eye image sensor: construction and calibration. In *Proc. SPIE*, volume 5301, pages 13–24, San Jose, CA, January 2004.

[318] K-.H. Jeong, J. Kim, and L.P. Lee. Biologically Inspired Artificial Compound Eyes. *Science*, 312:557–561, 2006.

[319] J. Tanida, R. Shogenji, Y. Kitamura, K. Yamada, M. Miyamoto, and S. Miyatake. Imaging with an integrated compound imaging system. *Opt. Express*, 11(18):2109–2117, September 2003.

[320] R. Shogenji, Y. Kitamura, K. Yamada, S. Miyatake, and J. Tanida. Multispectral imaging using compact compound optics. *Opt. Express*, 12(8):1643–1655, April 2004.

[321] S. Miyatake, R. Shogenji, M. Miyamoto, K. Nitta, and J. Tanida. Thin observation module by bound optics (TOMOBO) with color filters. In *Proc. SPIE*, volume 5301, pages 7–12, San Jose, CA, January 2004.

[322] R. Ng, M. Levoy, M. Brédif, G. Duval, M. Horowitz, and P. Hanrahan. Light Field Photography with a Hand-held Plenoptic Camera. *Stanford Tech Report*, CSTR-2005-02:1–11, 2005.

[323] A. Wang, P. Gill, and A. Molnar. An Angle-Sensitive CMOS Imager for Single-Sensor 3D Photography. In *Dig. Tech. Papers Int'l Solid-State Circuits Conf. (ISSCC)*, pages 412–414, 2011.

[324] A. Wang, P. Gill, and A. Molnar. Light field image sensors based on the Talbot effect. *Appl. Opt.*, 48(31):5897–5905, November 2009.

[325] A. Wang and A. Molnar. A Light-Field Image Sensor in 180 nm CMOS. *IEEE J. Solid-State Circuits*, 47(1):257–271, January 2012.

[326] S. Koyama, K. Onozawa, K. Tanaka, S. Saito, S.M. Kourkouss, and Y. Kato. Multiocular image sensor with on-chip beam-splitter and inner meta-micro-lens for single-main-lens stereo camera. *Opt. Express*, 24(16):18035–18048, August 2016.

[327] M. Sarkar and A. Theuwissen, editors. *A Biologically Inspired CMOS Image Sensor*. Springer, Berlin, Germany, 2013.

[328] Z.K. Kalayjian, A.G. Andreou, L. Wolff, and N. Sheppard. A Polarization Contrast Retina Using Patterned Iodine-doped PVA Film. In *Proc. European Solid-State Circuits Conf. (ESSCIRC)*, Neuchate, Switzerland, September 1996.

[329] V. Gruev, J. Van der Spiegel, and N. Engheta. Low Power Image Sensor With Polymer Polarization Filters. In *Int'l Symp. Circuits & Systems (ISCAS)*, Seattle, WA, USA, May 2008.

[330] Z.K. Kalayjian and A.G. Andreou. A silicon retina for polarization contrast vision. In *Int'l Joint Conf. Neural Networks (IJCNN)*, Washington, DC, USA, July 1999.

[331] M. Momeni and A.H.Titus. An Analog VLSI Chip Emulating Polarization Vision of *Octopus* Retina. *IEEE Trans. Neural Net.*, 17(1):222–232, January 2006.

[332] G. Myhre, W.-L. Hsu, A. Peinado, C. LaCasse, N. Brock, R.A. Chipman, and S. Pau. Liquid crystal polymer full-stokes division of focal plane polarimeter. *Opt. Express*, 20(25):27393–27409, December 2012.

[333] T. Sato, T. Araki, Y. Sasaki, T. Tsuru, T. Tadokoro, and S. Kawakami. Compact ellipsometer employing a static polarimeter module with arrayed polarizer and wave-plate elements. *Appl. Opt.*, 46(22):4963–4967, August 2007.

[334] T. Tokuda, S. Sato, H. Yamada, and J. Ohta. Polarization analyzing CMOS sensor for microchamber/microfluidic system based on image sensor technology. In *Int'l Symp. Circuits & Systems (ISCAS)*, Seattle, WA, USA, 2008.

[335] M. Sarkar, D.S.S. Bello, C.v. Hoof, and A.J.P. Theuwissen. Integrated Polarization Analyzing CMOS Image Sensor for Material Classification. *IEEE Sensors Journal*, 11(8):1692–1703, August 2011.

[336] M. Zhang, X. Wu, N. Cui, N. Engheta, and J. Van der Spiegel. Bioinspired Focal-Plane Polarization Image Sensor: Design From Application to Implementation. *Proc. IEEE*, 102(10):1435–1449, October 2014.

[337] Y. Maruyama, T. Terada, T. Yamazaki, Y. Uesaka, M. Nakamura, Y. Matoba, K. Komori, Y. Ohba, S. Arakawa, Y. Hirasawa, Y. Kondo, J. Murayama, K. Akiyama, Y. Oike, S. Sato, , and T. Ezaki. 3.2-MP Back-Illuminated Polarization Image Sensor With Four-Directional Air-Gap Wire Grid and 2.5-μm Pixels. *IEEE Trans. Electron Dev.*, 65(6):2544–2551, June 2018.

[338] K. Sasagawa, S. Shishido, K. Ando, H. Matsuoka, T. Noda, T. Tokuda, K. Kakiuchi, and J. Ohta. Image sensor pixel with on-chip high extinction ratio polarizer based on 65-nm standard CMOS technology. *Opt. Express*, 21(9):11132–11140, April 2013.

[339] T. Tokuda, K. Sasagawa, N. Wakama, T. Noda, K. Kakiuchi, and J. Ohta. Demonstrations of Polarization Imaging Capability and Novel Functionality of Polarization-Analyzing CMOS Image Sensor with 65 nm Standard CMOS Process. *ITE Trans. Media Tech. Appl.*, 2(2):131–138, 2014.

[340] Tomohiro Yamazaki, Yasushi Maruyama, Yusuke Uesaka, Motoaki Nakamura, Yoshihisa Matoba, Takashi Terada, Kenta Komori, Yoshiyuki Ohba, Shinichi Arakawa, Yasutaka Hirasawa, Yuhi Kondo, Jun Murayama, Kentaro Akiyama, Yusuke Oike, Shuzo Sato, and Takayuki Ezaki. Four-Directional Pixel-Wise Polarization CMOS Image Sensor Using Air-Gap Wire Grid on 2.5-μm Back-Illuminated Pixels. In *Tech. Dig. Int'l Electron Devices Meeting (IEDM)*, San Francisco, CA, USA, December 2016.

[341] M. Zhang, X. Wu, N. Engheta, and J. Van der Spiegel. A Monolithic CMOS Image Sensor With Wire-Grid Polarizer Filter Mosaic in the Focal Plane. *IEEE Trans. Electron Dev.*, 61(3):855–862, March 2014.

[342] M. Sarkar, D.S.S. Bello, C.v. Hoof, and A.J.P. Theuwissen. Low Power Image Sensor With Polymer Polarization Filters. In *Int'l Symp. Circuits & Systems (ISCAS)*, pages 621 – 624, Paris, France, May 2010.

[343] V. Gruev, R. Perkins, and T. York. CCD polarization imaging sensor with aluminum nanowire optical filters. *Opt. Express*, 18(18):19087–19094, August 2010.

[344] T. York, S.B. Powell, S. Gao, L. Kahan, T. Charanya, D. Saha, N.W. Roberts, T.W. Cronin, J. Marshall, S. Achilefu, S.P. Lake, B. Raman, and V. Gruev. Bioinspired Polarization Imaging Sensors: From Circuits and Optics to Signal Processing Algorithms and Biomedical Applications. *Proc. IEEE*, 102(10):1450–1469, October 2014.

[345] V. Boominathan, J.K. Adams, M.S. Asif, B. Avants, J.T. Robinson, R.G. Baraniuk, A.C. Sankaranarayanan, and A. Veeraraghavan. Lensless Imaging:

A Computational Renaissance. *IEEE Sig. Proc. Mag.*, 33(5):23–35, September 2016.

[346] E. E. Fenimore and T. M. Cannon. Coded aperture imaging with uniformly redundant arrays. *Appl. Opt.*, 17(3):337–347, February 1978.

[347] T.M. Cannon and E.E. Feinmore. Coded aperture imaging: many holes make light work. *Opt. Eng.*, 19(3):283–289, May 1980.

[348] H.H. Barrett. Coded aperture imaging: many holes make light work. *J. Nucl. Med.*, 13(6):382–385, June 1972.

[349] P.R. Gill, C. Lee, D.-G. Lee, A. Wang, and A. Molnar. A microscale camera using direct Fourier-domain scene capture. *Opt. Lett.*, 36(15):2949–2951, August 2011.

[350] D.G. Stork and P.R. Gill. Lensless Ultra-Miniature CMOS Computational Imagers and Sensors. In *Int'l Conf. Sensor Technologies and Applications (SENSORCOMM)*, Barcelona, Spain, August 2013.

[351] M.S. Asif, A. Ayremlou, A. Sankaranarayanan, A. Veeraraghavan, and R. Baraniuk. FlatCam: Thin, Bare-Sensor Cameras using Coded Aperture and Computation. *arXiv*, 1509.00116v2, January 2016.

[352] Y. Nakamura, T. Shimano, K. Tajima, M. Sao, and T. Hoshizawa. Lensless Light-field Imaging with Fresnel Zone Aperture. In *Int'l Workshop on Image Sensor and Systems (IWISS2016)*, Tokyo, Japan, 2016.

[353] A. Zomet and S.K. Nayar. Lensless Imaging with a Controllable Aperture. In *IEEE Computer Society Conf. Computer Vision and Pattern Recognition (CVPR'06)*, New York, NY, USA, USA, June 2006.

[354] G. Huang, H. Jiang, K. Matthews, and P. Wilford. Lensless imaging by compressive sensing. In *IEEE Int'l Conf. Image Processing*, Melbourne, VIC, Australia, September 2012.

[355] M.J. DeWeert and B.P. Farm. Lensless coded-aperture imaging with separable Doubly-Toeplitz masks. *Opt. Eng.*, 54(2):023102–1–9, February 2015.

[356] A. Greenbaum, W. Luo, T. Su, Z. Göröcs, L. Xue, S.O Isikman, A.F Coskun, O. Mudanyali, and A Ozcan. Imaging without lenses: achievements and remaining challenges of wide-field on-chip microscopy. *Nat. Method*, 9(9):889–895, September 2012.

[357] A. Ozcan and E. McLeod. Lensless Imaging and Sensing. *Annu. Rev. Biomed. Eng.*, 18:77–102, January 2016.

[358] M. Roy, D. Seo, S. Oh, J. Yang, and S. Seo. A review of recent progress in lens-free imaging and sensing. *Biosensors & Bioelectron.*, 88:130–143, February 2017.

[359] X. Huang, J. Guo, X. Wang, M. Yan, Y. Kang, and H. Yu. A Contact-Imaging Based Microfluidic Cytometer with Machine-Learning for Single-Frame Super-Resolution Processing. *PLOS ONE*, 9(8):e104539, August 2014.

[360] Y. Fang, N. Yu, Y. Jiang, and C. Dang. High-Precision Lens-Less Flow Cytometer on a Chip. *Micromachines*, 9(277), May 2018.

[361] K. Sasagawa, S.H. Kim, K. Miyazawa, H. Takehara, T. Noda, T. Tokuda, R. Iino, H. Noji, and J. Ohta. Lensless CMOS-Based Imaging Device for Fluorecent Femtoliter Droplet Array Counting. In *Int'l Conf. Miniaturized Systems for Chemistry and Life Sciences (μTAS)*, Freiburg, Germany, 2013.

[362] K. Sasagawa, A. Kimura, M. Haruta, T. Noda, T. Tokuda, and J. Ohta. Highly sensitive lens-free fluorescence imaging device enabled by a complementary combination of interference and absorption filters. *Biomed. Opt. Express*, 9(9):4329–4344, September 2018.

[363] K. Imai1, M. Nishigaki, Y. Onozuka, Y. Akimoto, M. Nagai, S. Matsumoto, and S. Kousai. A lens-free single-shot fluorescent imaging system using CMOS image sensors with dielectric multi-layer filter. In *Int'l Conf. Solid-State Sensors, Actuators and Microsystems (TRANSDUCERS)*, Kaohsiung, Taiwan, June 2017.

[364] A.F. Coskun, I. Senca, T. Su, and A. Ozcan. Lensless wide-field fluorescent imaging on a chip using compressive decoding of sparse objects. *Opt. Express*, 18(10):10510–10523, May 2010.

[365] L. Hong, H. Li, H. Yang, and K. Sengupta. Fully Integrated Fluorescence Biosensors On-Chip Employing Multi-Functional Nanoplasmonic Optical Structures in CMOS. *IEEE J. Solid-State Circuits*, 52(9):2388–2046, September 2017.

[366] Hironari Takehara, Y. Nakamoto, N. Ikeda, K. Sasagawa, M. Haruta, T. Noda, T. Tokuda, and J. Ohta. Compact Lensless Fluorescence Counting System for Single Molecular Assay. *IEEE Trans. Biomedical Circuits & Sytems*, 12(5):1177 – 1185, May 2018.

[367] K. Sasagawa, Y. Ohta, M. Kawahara, M. Haruta, T. Tokuda, and J. Ohta. Wide field-of-view lensless fluorescence imaging device with hybrid bandpass emission filter. *AIP Adv.*, 9:035108, 2019.

[368] K. Bengler, K. Dietmayer, B. Färber, M. Maurer, C. Stiller, and H. Winner. Three Decades of Driver Assistance Systems: Review and Future Perspectives. *IEEE Intelligent Transportation Systems Magazine*, 6(4):6–22, 2014.

[369] B.J. Hosticka, W. Brockherde, A. Bussmann, T. Heimann, R. Jeremias, A. Kemna, C. Nitta, and O. Schrey. CMOS imaging for automotive applications. *IEEE Trans. Electron Dev.*, 2003.

[370] D.J. Denvir and E. Conroy. Electron Multiplying CCD Technology: The new ICCD. In *Proc. SPIE*, volume 4796, pages 164–174, Seattle, WA, 2003.

[371] H. Eltoukhy, K. Salama, and A.E. Gamal. A 0.18-μm CMOS bioluminescence detection lab-on-chip. *IEEE J. Solid-State Circuits*, 41(3):651 – 662, March 2006.

[372] H. Jia and P.A. Abshire. A CMOS image sensor for low light applications. In *Int'l Symp. Circuits & Systems (ISCAS)*, pages 1651–1654, Kos, Greece, May 2006.

[373] M.-W. Seo, S. Kawahito, K. Kagawa, and K. Yasutomi. A 0.27e$_{rms}^{-}$ Read Noise 220-μV/e^{-} Conversion Gain Reset-Gate-Less CMOS Image Sensor With 0.11-μm CIS Process. *IEEE Electron Device Lett.*, 36(12):1344–1347, December 2015.

[374] S. Kawahito and M.-W. Seo. Noise Reduction Effect of Multiple-Sampling-Based Signal-Readout Circuits for Ultra-Low Noise CMOS Image Sensors. *Sensors*, 16(11):1867–1885, November 2016.

[375] A.M. Fowler and I. Gatley. Noise reduction strategy for hybrid IR focal-plane arrays. In *Proc. SPIE*, volume 1541, pages 127–133, November 1991.

[376] G. Lutz, M. Porro, S. Aschauer, S. Wölfel, and L. Struder. The DEPFET Sensor-Amplifier Structure: A Method to Beat 1/f Noise and Reach Sub-Electron Noise in Pixel Detectors. *Sensors*, 16(5):608–622, 2015.

[377] M. Schanz, C. Nitta, A. Bußmann, B.J. Hosticka, and R.K. Wertheimer. A high-dynamic-range CMOS image sensor for automotive applications. *IEEE J. Solid-State Circuits*, 35(7):932 – 938, September 2000.

[378] S. Kawahito, M. Sakakibara, D. Handoko, N. Nakamura, H. Satoh, M. Higashi, K. Mabuchi, and H. Sumi. A Column-Based Pixel-Gain-Adaptive CMOS Image Sensor for Low-Light-Level Imaging. In *Dig. Tech. Papers Int'l Solid-State Circuits Conf. (ISSCC)*, San Francisco, CA, February 200.

[379] M. Sakakibara, S. Kawahito, D. Handoko, N. Nakamura, H. Satoh, M. Higashi, K. Mabuchi, and H. Sumi. A high-sensitivity CMOS image sensor with gain-adaptive column amplifiers. *IEEE J. Solid-State Circuits*, 40(5):1147 – 1156, May 2005.

[380] Y. Nitta, Y. Muramatsu, K. Amano, T. Toyama, J. Yamamoto, K. Mishina, A. Suzuki, T. Taura, A. Kato, M. Kikuchi, Y. Yasui, H. Nomura, and N. Fukushima. High-Speed Digital Double Sampling with Analog CDS on Column Parallel ADC Architecture for Low-Noise Active Pixel Sensor. In *Dig. Tech. Papers Int'l Solid-State Circuits Conf. (ISSCC)*, pages 2024 – 2031, February 2006.

[381] D. Kim, J. Bae, and M. Song. A High Speed CMOS Image Sensor with a Novel Digital Correlated Double Sampling and a Differential Difference Amplifier. *Sensors*, 15:5081–5095, 2015.

[382] M. Furuta, T. Inoue, Y. Nishikawa, and S. Kawahito. A 3500fps High-Speed CMOS Image Sensor with 12b Column-Parallel Cyclic A/D Converters. In *Dig. Tech. Papers Symp. VLSI Circuits*, pages 21–22, June 2006.

[383] R. Funatsu, S. Huang, T. Yamashita, K. Stevulak, J. Rysinski, D. Estrada, S. Yan, T. Soeno, T. Nakamura, T. Hayashida, H. Shimamoto, and B. Mansoorian. 133Mpixel 60fps CMOS Image Sensor with 32-Column Shared High-Speed Column-Parallel SAR ADCs. In *Dig. Tech. Papers Int'l Solid-State Circuits Conf. (ISSCC)*, pages 112–113, 2015.

[384] H. Wakabayashi, A. Suzuki, T. Kainuma, C. Okada, N. Kawazu, T. Oka, Y. Yagasaki, S. Gonoi, and M. Mizuno. A 1/1.7-inch 20Mpixel Back-Illuminated Stacked CMOS Image Sensor with Multi-Functional Modes. *ITE Trans. Media Tech. Appl.*, 4(2):136–141, February 2016.

[385] S. Lauxtermann, A. Lee, J. Stevens, and A. Joshi. Comparison of Global Shutter Pixels for CMOS Image Sensors. In *Int'l Image Sensor Workshop*, pages 82–85, Ogunquit, Maine, USA, June 2007.

[386] T. Yokoyama, M. Tsutsui, M. Suzuki, Y. Nishi, I. Mizuno, and A. Lahav. Development of Low Parasitic Light Sensitivity and Low Dark Current 2.8 μm Global Shutter Pixel. *Sensors*, 18:349–361, January 2018.

[387] S. Velichko, J.J. Hynece, R.S. Johnson, V. Lenchenkov, H. Komori, H. Lee, and F.Y.J. Chen. CMOS Global Shutter Charge Storage Pixels With Improved Performance. *IEEE Trans. Electron Dev.*, 63(1):106–112, January 2016.

[388] A. Lahav, A. Birman, M. Cohen, T. Leitner, and A. Fenigstein. Design of photo-electron barrier for the Memory Node of a Global Shutter pixel based on a Pinned Photodiode. In *Int'l Image Sensor Workshop*, Bergen, Norway, June 2009.

[389] A. Lahav, A. Birman, D. Veinger, A. Fenigstein, D. Zhang, and D. van Blerkom. IR Enhanced Global Shutter Pixel for High Speed Applications. In *Int'l Image Sensor Workshop*, Snowbird, UT, USA, June 2013.

[390] H. Sekine, M. Kobayashi, Y. Onuki, K. Kawabata, T. TsuboiY. Matsuno, H. Takahashi, S. Inoue, and T. Ichikawa. A High Optical Performance 3.4 μm Pixel Pitch Global Shutter CMOS Image Sensor with Light Guide Structure. In *Int'l Image Sensor Workshop*, Hiroshima, Japan, June 2017.

[391] G. Meynants, B. Wolfs, J. Bogaerts, P. Li, Z. Li, Y. Li, Y. Creten, K. Ruythooren, P. Francis, R. Lafaille, P. De Wit, G. Beeckman, and J.M. Kopfer. A 47 MPixel 36.4 × 27.6 mm^2 30 fps Global Shutter Image Sensor. In *Int'l Image Sensor Workshop*, Hiroshima, Japan, June 2017.

[392] K. Kawabata, M. Kobayashi, Y. Onuki, H. Sekine, T. Tsuboi, Y. Matsuno, H. Takahashi, S. Inoue, and T. Ichikawa. A 1.8e$^-$ Temporal Noise Over 90dB Dynamic Range 4k2k Super 35mm format. In *Tech. Dig. Int'l Electron Devices Meeting (IEDM)*, San Francisco, CA, USA, December 2016.

[393] K. Yasutomi, Y. Sadanaga, T. Takasawa, S. Itoh, and S. Kawahito. Dark Current Characterization of CMOS Global Shutter Pixels Using Pinned Storage Diodes. In *Int'l Image Sensor Workshop*, Hokkaido, Japan, June 2011.

[394] M. Kobayashi, Y. Onuki, K. Kawabata, H. Sekine, T. Tsuboi, Y. Matsuno, H. Takahashi, T. Koizumi, K. Sakurai, H. Yuzurihara, S. Inoue, and T. Ichikawa. A $1.8e^-_{rms}$ Temporal Noise Over 110dB Dynamic Range 3.4 μm Pixel Pitch Global Shutter CMOS Image Sensor with Dual-Gain Amplifiers, SS-ADC and Multiple-Accumulation Shutter. In *Dig. Tech. Papers Int'l Solid-State Circuits Conf. (ISSCC)*, San Francisco, CA, USA, February 2017.

[395] P. Centen, S. Lehr, S. Roth, J. Rotte, F. Heizmann, A. Momin, R. Dohmen, K.-H. Schaaf, K. Jan Damstra, R. van Ree, and M. Schreiber. A 4e-noise 2/3-inch global shutter 1920x1080P120 CMOS-Imager. In *Int'l Image Sensor Workshop*, Snowbird, Utah, USA, June 2013.

[396] S. Velichko, G. Agranov, J. Hynecek, S. Johnson, H. Komori, J. Bai, I. Karasev, R. Mauritzson, X. Yi, V. Lenchenkov, S. Zhao, and H. Kim. Low Noise High Efficiency 3.75 μm and 2.8 μm Global Shutter CMOS Pixel Arrays. In *Int'l Image Sensor Workshop*, Snowbird, Utah, USA, June 2013.

[397] A. Lahav, D. Veinger, A. Birman, M. Suzuki, T. Hirata, K. Tachikawa, M. Tsutsui, T. Yokoyama, Y. Nishi, and I. Mizuno. Cross Talk, Quantum Efficiency, and Parasitic Light Sensitivity comparison for different Near Infra-Red enhanced sub 3μm Global Shutter pixel architectures. In *Int'l Image Sensor Workshop*, Hiroshima, Japan, June 2017.

[398] A. Toyoda, Y. Suzuki, K. Orihara, and Yasuaki Hokari. A Novel Tungsten Light-Shield Structure for High-Density CCD Image Sensors. *IEEE Trans. Electron Dev.*, 38(5):965–968, May 1991.

[399] M. Furuta, T. Inoue, Y. Nishikawa, and S. Kawahito. A 3500fps High-speed CMOS Image Sensor with 12b Column-Parallel Cyclic A/D Converters. In *Dig. Tech. Papers Int'l Solid-State Circuits Conf. (ISSCC)*, February 2006.

[400] T. Toyama, K. Mishina, H. Tsuchiya, T. Ichikawa, H. Iwaki, Y. Gendai, H. Murakami, K. Takamiya, H. Shiroshita, Y. Muramatsu, and T. Furusawa. A 17.7Mpixel 120fps CMOS image sensor with 34.8Gb/s readout. In *Dig. Tech. Papers Int'l Solid-State Circuits Conf. (ISSCC)*, February 2011.

[401] K. Kitamura, T. Watabe, T. Sawamoto, T. Kosugi, T. Akahori, T. Iida, K. Isobe, T. Watanabe, H. Shimamoto, H. Ohtake, S. Aoyama, S. Kawahito, and N. Egami. A 33-Megapixel 120-Frames-Per-Second 2.5-Watt CMOS Image Sensor With Column-Parallel Two-Stage Cyclic Analog-to-Digital Converters. *IEEE Trans. Electron Dev.*, 59(12):3246–3433, December 2012.

[402] I. Takayanagi and J. Nakamura. High-Resolution CMOS Video Image Sensors. *Proc. IEEE*, 101(1):61–73, January 2013.

[403] H. Honda, S. Osawa, M. Shoda, E. Pages, T. Sato, N. Karasawa, B. Leichner, J. Schoper, E.S. Gattuso, D. Pates, J. Brooks, S. Johnson, and I. Takayanagi. A 1-inch Optical Format, 14.2M-pixel, 80fps CMOS Image Sensor with a Pipelined Pixel Reset and Readout Operation. In *Dig. Tech. Papers Symp. VLSI Circuits*, June 2013.

[404] T. Yasue, K. Kitamura, T. Watabe, H. Shimamoto, T. Kosugi, T. Watanabe, S. Aoyama, M. Monoi, Z. Wei, and S. Kawahito. A 1.7-in, 33-Mpixel, 120-frames/s CMOS Image Sensor With Depletion-Mode MOS Capacitor-Based 14-b Two-Stage Cyclic A/D Converters. *IEEE Trans. Electron Dev.*, 63(1):153–161, January 2016.

[405] S. Sugawa. Ultra-high speed video imaging technologies. *ITE Tech. Report*, 37(19):9–14, March 2013. In Japanese.

[406] M. El-Desouki, M.J. Deen, Q. Fang, L. Liu, F. Tse, and D. Armstrong. CMOS Image Sensors for High Speed Applications. *Sensors*, 9:430–444, January 2009.

[407] M.M. El-Desouki, O. Marinov, M.J. Deen, and Q. Fang. CMOS Active-Pixel Sensor With In-Situ Memory for Ultrahigh-Speed Imaging. *IEEE Sensors Journal*, 11(6):1375–1379, June 2011.

[408] J.A. Schmitz, M.K. Gharzai, S. Balkır, M.W. Hoffman, D.J. White, and N. Schemm. A 1000 frames/s Vision Chip Using Scalable Pixel-Neighborhood-Level Parallel Processing. *IEEE J. Solid-State Circuits*, 52(2):556–568, February 2017.

[409] T.G. Etoh, D. Poggemann, G. Kreider, H. Mutoh, A.J.P. Theuwissen, A. Ruckelshausen, Y. Kondo, H. Maruno, K. Takubo, H. Soya, K. Takehara, T. Okinaka, and Y. Takano. An image sensor which captures 100 consecutive frames at 1000000 frames/s. *IEEE Trans. Electron Dev.*, 50(1):144–151, January 2003.

[410] T. Arai, J. Yonai, T. Hayashida, H. Ohtake, H. van Kuijk, and T.G. Etoh. A 252-V/lux·s, 16.7-Million-Frames-Per-Second 312-kpixel Back-Side-Illuminated Ultrahigh-Speed Charge-Coupled Device. *IEEE Trans. Electron Dev.*, 60(10):3540–3548, October 2013.

[411] T. G. Etoh, V. T. S. Dao, K. Shimonomura, E. Charbon, C. Zhang, Y. Kamakura, and T. Matsuoka. Toward 1Gfps: Evolution of ultra-high-speed image sensors -ISIS, BSI, multi-collection gates, and 3D-stacking-. In *Tech. Dig. Int'l Electron Devices Meeting (IEDM)*, San Francisco, CA, USA, December 2014.

[412] Y. Tochigi, K. Hanazawa, Y. Kato, R. Kuroda, H. Mutoh, R. Hirose, H. Tominaga, K. Takubo, Y. Kondo, and S. Sugawa. A Global-Shutter CMOS Image Sensor With Readout Speed of 1-Tpixel/s Burst and 780-Mpixel/s Continuous. *IEEE J. Solid-State Circuits*, 48(1):329–338, January 2013.

[413] K. Miyauchi, T. Takeda, K. Hanzawa, Y. Tochigi, S. Sakai, R. Kuroda, H. Tominaga, R. Hirose, K. Takubo, Y. Kondo, and S. Sugawa. Pixel Structure with 10 nsec Fully Charge Transfer Time for the 20M Frame Per Second Burst CMOS Image Sensor. In *Proc. SPIE*, volume 9022, pages 902203–1–12, 2014.

[414] R. Kuroda, Y. Tochigi, K. Miyauchi, T. Takeda, H. Sugo, F. Shao, and S. Sugawa. A 20Mfps Global Shutter CMOS Image Sensor with Improved Light Sensitivity and Power Consumption Performances. *ITE Trans. Media Tech. Appl.*, 4(2):149–154, February 2016.

[415] C. Tubert, L. Simony, F. Roy, A. Tournier, L. Pinzelli, and P. Magnan. High Speed Dual Port Pinned-photodiode for Time-Of-Flight Imaging. In *Int'l Image Sensor Workshop*, Bergen, Norway, June 2009.

[416] D.H. Hubel. *Eye, Brain, and Vision*. Scientific American Library, New York, NY, 1987.

[417] M. Mase, S. Kawahito, M. Sasaki, Y. Wakamori, and M. Furuta. A Wide Dynamic Range CMOS Image Sensor With Multiple Exposure-Time Signal Outputs and 12-bit Column-Parallel Cyclic AD Converters. *IEEE J. Solid-State Circuits*, 40(12):2787–2795, December 2005.

[418] S. Iida, Y. Sakano, T. Asatsuma, M. Takami, I. Yoshiba, N. Ohba, H. Mizuno, T. Oka, K. Yamaguchi, A. Suzuki, K. Suzuki, M. Yamada, M. Takizawa, Y. Tateshita, and K. Ohno. A 0.68e-rms Random-Noise 121dB Dynamic-Range Sub-pixel architecture CMOS Image Sensor with LED Flicker Mitigation. In *Tech. Dig. Int'l Electron Devices Meeting (IEDM)*, San Francisco, CA, USA, December 2018.

[419] M. Innocent, A. Rodriguez, D. Guruaribam, M. Rahman, M. Sulfridge, S. Borthakur, B. Gravelle, T. Goto, N. Dougherty, B. Desjardin, D. Sabo, M. Mlinar, and T. Geurts. Pixel with nested photo diodes and 120 dB single exposure dynamic range. In *Int'l Image Sensor Workshop*, Snowbird, Utah, USA, June 2019.

[420] S. Sugawa, N. Akahane, S. Adachi, K. Mori, T. Ishiuchi, and K. Mizobuchi. A 100 dB dyamic range CMOS image sensor using a lateral overflow integration capacitor. In *Dig. Tech. Papers Int'l Solid-State Circuits Conf. (ISSCC)*, pages 352–353, February 2005.

[421] N. Akahane, R. Ryuzaki, S. Adachi, K. Mizobuchi, and S. Sugawa. A 200dB Dynamic Range Iris-less CMOS Image Sensor with Lateral Overflow Integration Capacitor using Hybrid Voltage and Current Readout Operation. In *Dig. Tech. Papers Int'l Solid-State Circuits Conf. (ISSCC)*, February 2006.

[422] N. Akahane, S. Sugawa, S. Adachi, K. Mori, T. Ishiuchi, and K. Mizobuchi. A sensitivity and linearity improvement of a 100-dB dynamic range CMOS image sensor using a lateral overflow integration capacitor. *IEEE J. Solid-State Circuits*, 41(4):851 – 858, April 2006.

[423] S.Kavadias, B. Dierickx, D. Scheffer, A. Alaerts, D. Uwaerts, and J. Bogaerts. A logarithmic response CMOS image sensor with on-chip calibration. *IEEE J. Solid-State Circuits*, 35(8):1146–1152, August 2000.

[424] E.C. Fox, J. Hynecek, and D.R. Dykaar. Wide-Dynamic-Range Pixel with Combined Linear and Logarithmic Response and Increased Signal Swing. In *Proc. SPIE*, volume 3965, pages 4–10, 2000.

[425] G.G. Storm, J.E.D. Hurwitz, D. Renshawa, K.M. Findlater, R.K. Henderson, and M.D. Purcell. High dynamic range imaging using combined linear-logarithmic response from a CMOS image sensor. In *IEEE Workshop on Charge-Coupled Devices & Advanced Image Sensors*, Elmau, Germany, May 2003.

[426] K. Hara, H. Kubo, M. Kimura, F. Murao, and S. Komori. A Linear-Logarithmic CMOS Sensor with Offset Calibration Using an Injected Charge Signal. In *Dig. Tech. Papers Int'l Solid-State Circuits Conf. (ISSCC)*, 2005.

[427] S. Decker, D. McGrath, K. Brehmer, and C.G.Sodini. A 256 × 256 CMOS imaging array with wide dynamic range pixels and column-parallel digital output. *IEEE J. Solid-State Circuits*, 33(12):2081–2091, December 1998.

[428] Y. Muramatsu, S. Kurosawa, M. Furumiya, H. Ohkubo, and Y. Nakashiba. A Signal-Processing CMOS Image Sensor using Simple Analog Operation. In *Dig. Tech. Papers Int'l Solid-State Circuits Conf. (ISSCC)*, pages 98–99, February 2001.

[429] V. Berezin, I. Ovsiannikov, D. Jerdev, and R. Tsai. Dynamic Range Enlargement in CMOS Imagers with Buried Photodiode. In *IEEE Workshop on Charge-Coupled Devices & Advanced Image Sensors*, Elmau, Germany, 2003.

[430] N. Bock, A. Krymski, A. Sarwari, M. Sutanu, N. Tu, K. Hunt, M. Cleary, N. Khaliullin, and M. Brading. A wide-VGA CMOS image sensor with global shutter and extended dyamic range. In *IEEE Workshop on Charge-Coupled Devices & Advanced Image Sensors*, pages 222–225, Karuizawa, Japan, June 2005.

[431] K. Kagawa, Y. Adachi, Y. Nose, H. Takashima, K. Tani, A. Wada, M. Nunoshita, and J. Ohta. A wide-dynamic-range CMOS imager by hybrid use of active and passive pixel sensors. In *IS&T SPIE Annual Symposium Electronic Imaging*, [6501-18], San Jose, CA, January 2007.

[432] S.T. Smith, P. Zalud, J. Kalinowski, N.J. McCaffrey, P.A. Levine, and M.L. Lin. BLINC: a 640 × 480 CMOS active pixel vide camera with adaptive digital processing, extended optical dynamic range, and miniature form factor. In *Proc. SPIE*, volume 4306, pages 41–49, January 2001.

[433] H. Witter, T. Walschap, G. Vanstraelen, G. Chapinal, G. Meynants, and B. Dierickx. 1024 × 1280 pixel dual shutter APS for industrial vision. In *Proc. SPIE*, volume 5017, pages 19–23, 2003.

[434] O. Yadid-Pecht and E.R. Fossum. Wide intrascene dynamic range CMOS APS using dual sampling. *IEEE Trans. Electron Dev.*, 44(10):1721–1723, October 1997.

[435] O. Schrey, J. Huppertz, G. Filimonovic, A. Bussmann, W. Brockherde, and B.J. Hosticka. A 1 K × 1 K high dynamic range CMOS image sensor with on-chip programmable region-of-interest. *IEEE J. Solid-State Circuits*, 37(7):911–915, September 2002.

[436] K. Mabuchi, N. Nakamura, E. Funatsu, T. Abe, T. Umeda, T. Hoshino, R. Suzuki, and H. Sumi. CMOS image sensor using a floating diffusion driving buried photodiode. In *Dig. Tech. Papers Int'l Solid-State Circuits Conf. (ISSCC)*, pages 112 – 516, February 2004.

[437] M. Sasaki, M. Mase, S . Kawahito, and Y. Tadokoro. A Wide Dyamic Range CMOS Image Sensor wit Integration of Short-Exposure-Time Signals. In *IEEE Workshop on Charge-Coupled Devices & Advanced Image Sensors*, Elmau, Germany, 2003.

[438] M. Sasaki, M. Mase, S. Kawahito, and Y. Tadokoro. A wide dynamic range CMOS image sensor with multiple short-time exposures. In *Proc. IEEE Sensors*, volume 2, pages 967–972, October 2004.

[439] D. Yang, A. El Gamal, B. Fowler, and H. Tian. A 640 × 512 CMOS Image Sensor with Ultrawide Dynamic Range Floating-Point Pixel-Level ADC. *IEEE J. Solid-State Circuits*, 34(12):1821–1834, December 1999.

[440] T. Hamamoto and K. Aizawa. A computational image sensor with adaptive pixel-based integration time. *IEEE J. Solid-State Circuits*, 36(4):580–585, April 2001.

[441] T. Yasuda, T. Hamamoto, and K. Aizawa. Adaptive-integration-time image sensor with real-time reconstruction function. *IEEE Trans. Electron Dev.*, 50(1):111– 120, January 2003.

[442] O. Yadid-Pecht and A. Belenky. In-pixel autoexposure CMOS APS. *IEEE J. Solid-State Circuits*, 38(8):1425– 1428, August 2003.

[443] P.M. Acosta-Serafini, I. Masaki, and C.G. Sodini. A 1/3" VGA linear wide dynamic range CMOS image sensor implementing a predictive multiple sampling algorithm with overlapping integration intervals. *IEEE J. Solid-State Circuits*, 39(9):1487 – 1496, September 2004.

[444] T. Anaxagoras and N.M. Allinson. High dynamic range active pixel sensor. In *Proc. SPIE*, volume 5301, pages 149–160, San Jose, CA, January 2004.

[445] D. Stoppa, A. Simoni, L. Gonzo, M. Gottardi, and G.-F. Dalla Betta. Novel CMOS image sensor with a 132-dB dynamic range. *IEEE J. Solid-State Circuits*, 37(12):1846– 1852, December 2002.

[446] J. Döge, G. Shöneberg, G.T. Streil, and A. König. An HDR CMOS Image Sensor With Spiking Pixels, Pixel-Level ADC, and Linear Characteristics. *IEEE Trans. Circuits & Systems II*, 49(2):155–158, February 2002.

[447] S. Chen, A. Bermak, and F. Boussaid. A compact reconfigurable counter memory for spiking pixels. *IEEE Electron Device Lett.*, 27(4):255–257, April 2006.

[448] K. Oda, H. Kobayashi, K. Takemura, Y. Takeuchi, and T. Yamda. The development of wide dynamic range image sensor. *ITE Tech. Report*, 27(25):17–20, 2003. In Japanese.

[449] D. Yang and A. El Gamal. Comparative analysis of SNR for image sensors with widened dynamic range. In *Proc. SPIE*, volume 3649, pages 197–211, San Jose, CA, February 1999.

[450] Y. Wang, S.L. Barna, S. Campbell, and E.R. Fossum. A high dynamic range CMOS APS image sensor. In *IEEE Workshop on Charge-Coupled Devices & Advanced Image Sensors*, pages 137–140, Lake Tahoe, NV, June 2001.

[451] S.L. Barna, L.P. Ang, B. Mansoorian, and E.R. Fossum. A low-light to sunlight, 60 frames/s, 80 kpixel CMOS APS camera-on-a-chip with 8b digital output. In *IEEE Workshop on Charge-Coupled Devices & Advanced Image Sensors*, pages 148–150, Karuizawa, Japan, June 1999.

[452] S. Lee and K. Yang. High dynamic-range CMOS image sensor cell based on self-adaptive photosensing operation. *IEEE Trans. Electron Dev.*, 2006.

[453] S. Ando and A. Kimachi. Time-Domain Correlation Image Sensor: First CMOS Realization of Demodulation Pixels Array. In *IEEE Workshop on Charge-Coupled Devices & Advanced Image Sensors*, pages 33–36, Karuizawa, Japan, June 1999.

[454] R. Lange and P. Seitz. Solid-state time-of-flight range camera. *IEEE J. Quantum Electron.*, 37(3):390–397, March 2001.

[455] A. Kimachi, T. Kurihara, M. Takamoto, and S. Ando. A Novel Range Finding Systems Using Correlation Image Sensor. *Trans.IEE Jpn.*, 121-E(7):367 – 375, July 2001.

[456] K. Yamamoto, Y. Oya, K. Kagawa, J. Ohta, M. Nunoshita, and K. Watanabe. Demonstration of a freqency-demodulation CMOS image sensor and its improvement of image quality. In *IEEE Workshop on Charge-Coupled Devices & Advanced Image Sensors*, Elmau, Germany, June 2003.

[457] S. Ando and A. Kimachi. Correlation Image Sensor: Two-Dimensional Matched Detection of Amplitude-Modulated Light. *IEEE Trans. Electron Dev.*, 50(10):2059–2066, October 2003.

[458] J. Ohta, K. Yamamoto, T. Hirai, K. Kagawa, M. Nunoshita, M. Yamada, Y. Yamasaki, S. Sughishita, and K. Watanabe. An image sensor with an in-pixel demodulation function for detecting the intensity of a modulated light signal. *IEEE Trans. Electron Dev.*, 50(1):166– 172, January 2003.

[459] K. Yamamoto, Y. Oya, K. Kagawa, J. Ohta, M. Nunoshita, and K. Watanabe. Improvement of demodulated image quality in a demodulated image sensor. *J. Inst. Image Information & Television Eng.*, 57(9):1108–1114, September 2003. In Japanese.

[460] Y. Oike, M. Ikeda, and K. Asada. A 120 × 110 position sensor with the capability of sensitive and selective light detection in wide dynamic range for robust active range finding. *IEEE J. Solid-State Circuits*, 39(1):246 – 251, January 2004.

[461] Y. Oike, M. Ikeda, and K. Asada. Pixel-Level Color Demodulation Image Sensor for Support of Image Recognition. *ieicee*, 2004.

[462] K. Yamamoto, Y. Oya, K. Kagawa, M. Nunoshita, J. Ohta, and K. Watanabe. A 128 × 128 Pixel CMOS Image Sensor with an Improved Pixel Architecture for Detecting Modulated Light Signals. *Opt. Rev.*, 13(2):64–68, April 2006.

[463] B. Buttgen, F. Lustenberger, and P. Seitz. Demodulation Pixel Based on Static Drift Fields. *IEEE Trans. Electron Dev.*, 53(11):2741 – 2747, November 2006.

[464] R. Lange, P. Seitz, A. Biber, and S. Lauxtermann. Demodulation Pixels in CCD and CMOS Technologies for Time-Of-Flight Ranging. In *Proc. SPIE*, pages 177–188, San Jose, CA, January 2000.

[465] A. Kimachi, H. Ikuta, Y. Fujiwara, and H. Matsuyama. Spectral matching imager using correlation image sensor and AM-coded multispectral illumination. In *Proc. SPIE*, volume 5017, pages 128–135, Santa Clara, CA, January 2003.

[466] K. Asashi, M. Takahashi, K. Yamamoto, K. Kagawa, and J. Ohta. Application of a demodulated image sensos for a camera system suppressing saturation. *J. Inst. Image Information & Television Eng.*, 60(4):627–630, April 2006. In Japanese.

[467] Y. Oike, M. Ikeda, and K. Asada. Design and implementation of real-time 3-D image sensor with 640 × 480 pixel resolution. *IEEE J. Solid-State Circuits*, 39(4):622 – 628, April 2004.

[468] B. Aull, J. Burns, C. Chen, B. Felton, H. Hanson, C. Keast, J. Knecht, A. Loomis, M. Renzi, A. Soares, V. Suntharalingam, K. Warner, D. Wolfson, D. Yost, and D. Young. Laser Radar Imager Based on 3D Integration of Geiger-Mode Avalanche Photodiodes with Two SOI Timing Circuit Layers. In *Dig. Tech. Papers Int'l Solid-State Circuits Conf. (ISSCC)*, pages 1179–1188, February 2006.

[469] D. Stoppa, L. Viarani, A. Simoni, L. Gonzo, M. Malfatti, and G. Pedretti. A 50 × 50-pixel CMOS sensor for TOF-based real time 3D imaging. In *IEEE Workshop on Charge-Coupled Devices & Advanced Image Sensors*, pages 230–233, Karuizawa, Japan, June 2005.

[470] O. Elkhalili, O.M. Schrey, P. Mengel, M. Petermann, W. Brockherde, and B.J. Hosticka. A 4 × 64 pixel CMOS image sensor for 3-D measurement applications. *IEEE J. Solid-State Circuits*, 30(7):1208–1212, February 2004.

[471] S. Kawahito, I.A. Halin, T. Ushinaga, T. Sawada, M. Homma, and Y. Maeda. A CMOS Time-of-Flight Range Image Sensor With Gates-on-Field-Oxide Structure. *IEEE Sensors Journal*, 7(12):1578–1586, December 2007.

[472] T. Kahlmann, F. Remondino, and H. Ingensand. Calibration for increased accuracy of the range imaging camera SwissRanger^{TM}. In *Int'l Arch. Photogrammetry, Remote Sensing & Spatial Information Sci.*, volume XXXVI part 5, pages 136–141. Int'l Soc. Photogrammetry & Remote Sensing (ISPRS) Commission V Symposium, September 2006.

[473] S.B. Gokturk, H. Yalcin, and C. Bamji. A Time-Of-Flight Depth Sensor – System Description, Issues and Solutions. In *Conf. Computer Vision & Pattern Recognition Workshop (CVPR)*, pages 35 – 44, Washington, DC, June 2004.

[474] T. Kato, S. Kawahito, K. Kobayashi, H. Sasaki, T. Eki, and T. Hisanaga. A Binocular CMOS Range Image Sensor with Bit-Serial Block-Parallel Interface Using Cyclic Pipelined ADCs. In *Dig. Tech. Papers Symp. VLSI Circuits*, pages 270–271, Honolulu, Hawaii, June 2002.

[475] S. Kakehi, S. Nagao, and T. Hamamoto. Smart Image Sensor with Binocular PD Array for Tracking of a Moving Object and Depth Estimation. In *Int'l Symposium Intelligent Signal Processing & Communication Systems (ISPACS)*, pages 635–638, Awaji, Japan, December 2003.

[476] R.M. Philipp and R. Etienne-Cummings. A 128 × 128 33mW 30frames/s single-chip stereo imager. In *Dig. Tech. Papers Int'l Solid-State Circuits Conf. (ISSCC)*, pages 2050 – 2059, February 2006.

[477] Y. Oike, M. Ikeda, and K. Asada. A 375 × 365 high-speed 3-D range-finding image sensor using row-parallel search architecture and multisampling technique. *IEEE J. Solid-State Circuits*, 40(2):444 – 453, February 2005.

[478] T. Sugiyama, S. Yoshimura, R. Suzuki, and H. Sumi. A 1/4-inch QVGA color imaging and 3-D sensing CMOS sensor with analog frame memory. In *Dig. Tech. Papers Int'l Solid-State Circuits Conf. (ISSCC)*, pages 434 – 479, February 2002.

[479] H. Miura, H. Ishiwata, Y. Lida, Y. Matunaga, S. Numazaki, A. Morisita, N. Umeki, and M. Doi. 100 frame/s CMOS active pixel sensor for 3D-gesture

recognition system. In *Dig. Tech. Papers Int'l Solid-State Circuits Conf. (ISSCC)*, pages 142 – 143, February 1999.

[480] M.D. Adams. Coaxial Range Measurement – Current Trends for Mobile Robotic Applications. *IEEE Sensors Journal*, 2(1):2 – 13, February 2002.

[481] L. Viarani, D. Stoppa, L. Gonzo, M. Gottardi, and A. Simoni. A CMOS Smart Pixel for Active 3-D Vision Applications. *IEEE Sensors Journal*, 4(1):145–152, February 2004.

[482] Y. Kato, T. Sano, Y. Moriyama, S. Maeda, T. Yamazaki, A. Nose, K. Shiina, Y. Yasu, W. van der Tempel, A. Ercan, Y. Ebiko, D. Van Nieuwenhove, and S. Sukegawa. 320×240 Back-Illuminated 10-μm CAPD Pixels for High-Speed Modulation Time-of-Flight CMOS Image Sensor. *IEEE J. Solid-State Circuits*, 53(4):1071–1078, April 2018.

[483] R. Jeremias, W. Brockherde, G. Doemens, B. Hosticka, L. Listl, and P. Mengel. A CMOS photosensor array for 3D imaging using pulsed laser. In *Dig. Tech. Papers Int'l Solid-State Circuits Conf. (ISSCC)*, pages 252 – 253, February 2001.

[484] T Ushinaga, I.A. Halin, T. Sawada, S. Kawahito, M. Homma, and Y. Maeda. A QVGA-size CMOS time-of-flight range image sensor with background light charge draining structure. In *Proc. SPIE*, pages 13–16, San Jose, CA, USA, 2006.

[485] T. Moller, H. Kraft, J. Frey, M. Albrecht, and R. Lange. Robust 3D Measurement with PMD Sensors. In *Proc. 1st Range Imaging Research Day at ETH Zurich*, page Supplement to the Proceedings, Zurich, 2005.

[486] J. Geng. Structured-light 3D surface imaging: a tutorial. *Advanced Optics and Photonics*, 3:129–160, 2011.

[487] Y. Oike, M. Ikeda, and K. Asada. A CMOS Image Sensor for High-Speed Active Range Finding Using Column-Parallel Time-Domain ADC and Position Encoder. *IEEE Trans. Electron Dev.*, 50(1):152–158, January 2003.

[488] A. Gruss, L.R. Carley, and T. Kanade. Integrated Sensor and Range-Finding Analog Signal Processor. *IEEE J. Solid-State Circuits*, 26(3):184–191, March 1991.

[489] D. Vallancourt and S.J. Daubert. Applications of current-copier circuits. In C. Toumazou, F.J. Lidgey, and D.G. Haigh, editors, *Analogue IC design: the current-mode approach*, IEE Circuits and Systems Series 2, chapter 14, pages 515–533. Peter Peregrinus Ltd., London, UK, 1990.

[490] M. Nakagawa. Ubiquitous Visible Light Communications. *IEICE Trans. on Communications*, J88-B(2):351–359, February 2005.

[491] M.S. Islim, R.X. Ferreira, X. He, E. Xie, S. Videv, S. Viola, S. Watson, N. Bamiedakis, R.V. Penty, I.H. White, A.E. Kelly, E. Gu, H. Haas, and M.D.

Dawson. Towards 10 Gb/s orthogonal frequency division multiplexing-based visible light communication using a GaN violet micro-LED. *Photonics Research*, 5(2):35–43, April 2017.

[492] J.R. Barry. *Wireless infrared communications*. Kluwer Academic Publishers, New York, NY, 1994.

[493] http://www.victor.co.jp/pro/lan/index.html.

[494] S. Rajagopal, R.D. Roberts, and S.-K. Lim. IEEE 802.15.7 Visible Light Communication: Modulation Schemes and Dimming Support. *IEEE Commun. Mag.*, 50(12):72–82, March 2012.

[495] T. Yamazato, I. Takai, H. Okada, T. Fujii, T. Yendo, S. Arai, M. Andoh, T. Harada, K. Yasutomi, K. Kagawa, and S. Kawahito. Image-Sensor-Based Visible Light Communication for Automotive Applications. *IEEE Commun. Mag.*, 52(7):88–97, July 2014.

[496] A. Jovicic, J. Li, and T. Richardson. Visible Light Communication: Opportunities, Challenges and the Path to Market. *IEEE Commun. Mag.*, 51(12):26–32, December 2013.

[497] L. Grobe, A. Paraskevopoulos, J. Hilt, D. Schulz, F.h Lassak, F. Hartlieb, C. Kottke, V. Jungnickel, and K.-D. Langer. High-Speed Visible Light Communication Systems. *IEEE Commun. Mag.*, 51(12):60–66, December 2013.

[498] D.C. O'Brien, G.E. Faulkner, E.B. Zyambo, K. Jim, D.J. Edwards, P. Stavrinou, G. Parry, J. Bellon, M.J. Sibley, V.A. Lalithambika, V.M. Joyner, R.J. Samsudin, D.M. Holburn, and R.J. Mears. Integrated Transceivers for Optical Wireless Communications. *IEEE Selected Topic Quantum Electron.*, 11(1):173–183, Jan-Feb. 2005.

[499] K. Kagawa, T. Nishimura, T.Hirai, Y. Yamasaki, J. Ohta, M. Nunoshita, and K. Watanabe. Proposal and Preliminary Experiments of Indoor optical wireless LAN based on a CMOS image sensor with a high-speed readout function enabling a low-power compact module with large downlink capacity. *IEICE Trans. Commun.*, E86-B(5):1498–1507, May 2003.

[500] K. Kagawa, H. Asazu, T. Kawakami, T. Ikeuchi, A. Fujiuchi, J. Ohta, M. Nunoshita, and K. Watanabe. Design and fabrication of a photoreceiver for a spatially optical communication using an image sensor. *J. Inst. Image Information & Television Eng.*, 58(3):334–343, March 2004. In Japanese.

[501] K. Kagawa, T. Ikeuchi, J. Ohta, and M. Nunoshita. An Image sensor with a photoreceiver function for indoor optical wireless LANs fabricated in 0.8-μm BiCMOS technology. In *Proc. IEEE Sensors*, page 288, Vienna, Austria, October 2004.

[502] K. Kagawa, J. Ohta, and J. Tanida. Dynamic Reconfiguration of Differential Pixel Output for CMOS Imager Dedicated to WDM-SDM Indoor Optical

Wireless LAN. *IEEE Photon. Tech. Lett.*, 21(18):1308–1310, September 2009.

[503] B.S. Leibowitz, B.E. Boser, and K.S.J. Pister. A 256-Element CMOS Imaging Receiver for Free-Space Optical Communication. *IEEE J. Solid-State Circuits*, 40(9):1948–1956, September 2005.

[504] M. Last, B.S. Leibowitz, B. Cagdaser, A. Jog, L. Zhou, B.E. Boser, and K.S.J. Pister. Toward a wireless optical communication link between two small unmanned aerial vehicles. In *Int'l Symp. Circuits & Systems (ISCAS)*, volume 3, pages 930–933, May 2003.

[505] S. Itoh, I. Takai, M.S.Z. Sarker, M. Hamai, K. Yasutomi, M. Andoh, and S. Kawahito. A CMOS Image Sensor for 10Mb/s 70m-Range LED Based Spatial Optical Communication. In *Dig. Tech. Papers Int'l Solid-State Circuits Conf. (ISSCC)*, pages 402–404, February 2010.

[506] I. Takai, S. Ito, K. Yasutomi, K. Kagawa, M. Andoh, and S. Kawahito. LED and CMOS Image Sensor Based Optical Wireless Communication System for Automotive Applications. *IEEE Photo. J.*, 5(5):6801418, October 2013.

[507] I. Takai, T. Harada, M. Andoh, K. Yasutomi, K. Kagawa, and S. Kawahito. Optical Vehicle-to-Vehicle Communication System Using LED Transmitter and Camera Receiver. *IEEE Photo. J.*, 6(5):7902513, October 2014.

[508] K. Kamakura. Image Sensors Meet LEDs. *IEICE Trans. Commun.*, E100-B(6):917–925, June 2017.

[509] D. J. Moore, R. Want, B. L. Harrison, A. Gujar, and K. Fishkin. Implementing Phicons: Combining Computer Vision with InfraRed Technology for Interactive Physical Icons. In *Proc. ACM Symposium on User Interface Software and Technology (UIST)*, pages 67–68, 1999.

[510] http://picalico.casio.com/ja/. In Japanese.

[511] N. Iizuka and M. Kikuchi. "Picapicamera", a Smartphone App. which Realized Visible Light Communications using a Camera. *J. Illum. Eng. Inst. Jpn*, 98(10):546–549, June 2014. In Japanese.

[512] N. Iizuka. Image Sensor Communication - Current Status and Future Perspectives. *IEICE Trans. Commun.*, E100-B(6):911–916, June 2017.

[513] http://www.fujitsu.com/jp/services/infrastructure/network/digital/flowsign-light/. In Japanese.

[514] K. Kuraki, K. Kato, and R. Tanaka. Technology for LED light-ing with embedded information on object. *FUJITSU J.*, 66(5):88–93, September 2015. In Japanese.

[515] https://panasonic.net/cns/LinkRay/.

[516] M. Oshima, H. Aoyama, K. Nakashima, and T. Maeda.   Image Sensor-based Visible Light Communication Technology. *Panasonic Tech. J.*, 61(2):118–123, November 2015. In Japanese.

[517] C. Danakis, M. Afgani, G. Povey, I. Underwood, and H. Haas. Using a CMOS camera sensor for visible light communication. In *IEEE Globecom Workshops*, Anaheim, CA, USA, March 2012.

[518] N. Matsushita, D. Hihara, T. Ushiro, S. Yoshimura, and J. Rekimoto. ID Cam; A smart camera for scene capturing and ID recognition. *J. Information Process. Soc. Jpn.*, 43(12):3664–3674, December 2002.

[519] N. Matsushita, D. Hihara, T. Ushiro, S. Yoshimura, J. Rekimoto, and Y. Yamamoto.   ID CAM: A Smart Camera for Scene Capturing and ID Recognition. *Proc. IEEE & ACM Int'l Sympo. Mixed & Augmented Reality*, page 227, 2003.

[520] K. Kagawa, Y. Maeda, K. Yamamoto, Y. Masaki, J. Ohta, and M. Nunoshita. Optical navigation: a ubiquitous visual remote-control station for home information appliances. In *Proc. Optics Japan*, pages 112–113, 2004. In Japanese.

[521] K.Kagawa, K. Yamamoto, Y. Maeda, Y. Miyake, H. Tanabe, Y. Masaki, M. Nunoshita, and J. Ohta.   "OptNavi," a Multi-Purpose Visual Remote Controller of Home Information Appliances Using a Custom CMOS Image Sensor. *Forum on Info. Tech. Lett.*, 4:229–232, April 2005. In Japanese.

[522] K. Kagawa, R. Danno, K. Yamamoto, Y. Maeda, Y. Miyake, H. Tanabe, Y. Masaki, M. Nunoshita, and J. Ohta. Demonstration of mobile visual remote controller "OptNavi" system using home network. *J. Inst. Image Information & Television Eng.*, 60(6):897–908, June 2006.

[523] Y. Oike, M. Ikeda, and K. Asada. A smart image sensor with high-speed feeble ID-beacon detection for augmented reality system. In *Proc. European Solid-State Circuits Conf.*, pages 125–128, September 2003.

[524] Y. Oike, M. Ikeda, and K. Asada. Smart Image Sensor with High-speed High-sensitivity ID Beacon Detection for Augmented Reality System. *J. Inst. Image Information & Television Eng.*, 58(6):835–841, June 2004. In Japanese.

[525] J. Deguchi, T. Yamagishi, H. Majima, N. Ozaki, K. Hiwada, M. Morimoto, T. Ashitani, and S. Kousai. A 1.4Mpixel CMOS image sensor with multiple row-rescan based data sampling for optical camera communication. In *IEEE Asian Solid-State Circuits Conf. (A-SSCC)*, KaoHsiung, Taiwan, November 2014.

[526] K. Yamamoto, Y. Maeda, Y. Masaki, K. Kagawa, M. Nunoshita, and J. Ohta. A CMOS image sensor for ID detection with high-speed readout of multiple region-of-interests. In *IEEE Workshop on Charge-Coupled Devices & Advanced Image Sensors*, pages 165–168, Karuizawa, Japan, June 2005.

[527] K. Yamamoto, K. Kagawa, Y. Maeda, Y. Miyake, H. Tanabe, Y. Masaki, M. Nunoshita, and J. Ohta. An Opt-Navi system using a custom CMOS image sensor with a function of reading multiple region-of-interests. *J. Inst. Image Information & Television Eng.*, 59(12):1830–1840, December 2005. In Japanese.

[528] http://www.irda.org/.

[529] http://www.bluetooth.com/.

[530] K. Yamamoto, Y. Maeda, Y. Masaki, K. Kagawa, M. Nunoshita, and J. Ohta. A CMOS image sensor with high-speed readout of multiple region-of-interests for an Opto-Navigation system. In *Proc. SPIE*, volume 5667, pages 90–97, San Jose, CA, January 2005.

[531] T. Tokuda, H. Matsuoka, N. Tachikawa nd N. Wakama, K. Terao, M. Shibata, T. Noda, K. Sasagawa, Y. Nishiyama, K. Kakiuchi, and J. Ohta. CMOS sensor-based miniaturised in-line dual-functional optical analyser for high-speed, in situ chirality monitoring. *Sensors & Actuators B*, 176:1032–1037, January 2013.

[532] D.G. Hafemann, J.W. Parce, and H.M. McConnell. Light-Addressable Potentiometric Sensor for Biochemical Systems. *Science*, 240(4856):1182–1185, May 1988.

[533] M. Nakano, T. Yoshinobu, and H. Iwasaki. Scanning-laser-beam semiconductor pH-imaging sensor. *Sensors & Actuators B*, 20:119–123, 1994.

[534] T. Yoshinobu, H. Iwasaki, M. Nakano, S. Nomura, T. Nakanishi, S. Takamuatsu, and K. Tomita. Application of Chemical Imaging Sensors to Electro Generated pH Disribution. *Jpn. J. Appl. Phys.*, 37(3B):L353–L355, March 1988.

[535] H. Tanaka, T. Yoshinobu, and H. Iwasaki. Application of the chemical imaging sensor to electrophysiological measurement of a neural cell. *Sensors & Actuators B*, 59:21–25, 1999.

[536] T. Yoshinobu and H. Iwasaki and Y. Ui and K. Furuichi and Y. Ermolenko and Y. Mourzina and T. Waginer and N. Näther and M.J. Schöning. The light-addressable potentiometric sensor for multi-ion sensing and imaging. *Methods*, 37:94–102, 2005.

[537] P. Bergveld. Development of an Ion-Sensitive Solid-State Device for Neurophysiological Measurements. *IEEE Trans. Biomedical Eng.*, BME-17(1):70–71, January 1970.

[538] P. Bergveld. Development, Operation, and Application of the Ion-Sensitive Field-Effect Transistor as a Tool for Electrophysiology. *IEEE Trans. Biomedical Eng.*, BME-19(5):342–351, September 1972.

[539] T. Matsuo and M. Esashi. Methods of ISFET Fabrication. *Sensors & Actuators*, 1:77–96, 1981.

[540] K. Sawada, S. Mimura, K. Tomita, T. Nakanishi, H. Tanabe, M. Ishida, and T. Ando. Novel CCD-Based pH Imaging Sensor. *IEEE Trans. Electron Dev.*, 46(9):1346 – 1349, September 1999.

[541] K. Sawada, T. Shimada, T. Ohshina, H. Takao, and M. Ishida. Highly sensitive ion sensors using charge transfer technique. *Sensors & Actuators B*, 98(1):69–72, March 2004.

[542] B. Eversmann, M. Jenkner, F. Hofmann, C. Paulus, R. Brederlow, B. Holzapfl, P. Fromherz, M. Merz, M. Brenner, M. Schreiter, R. Gabl, K. Plehnert, M. Steinhauser, G. Eckstein, D. Schmitt-Landsiedel, and R. Thewes. A 128 × 128 CMOS biosensor array for extracellular recording of neural activity. *IEEE J. Solid-State Circuits*, 38(12):2306–2317, December 2003.

[543] U. Lu, B. Hu, Y. Shih, C.Wu, and Y. Yang. The design of a novel complementary metal oxide semiconductor detection system for biochemical luminescence. *Biosensors Bioelectron.*, 19(10):1185–1191, 2004.

[544] H. Ji, P.A. Abshire, M. Urdaneta, and E. Smela. CMOS contact imager for monitoring cultured cells. In *Int'l Symp. Circuits & Systems (ISCAS)*, pages 3491–3495, Kobe, Japan, May 2005.

[545] K. Sawada, T. Ohshina, T. Hizawa, H. Takao, and M. Ishida. A novel fused sensor for photo- and ion-sensing. *Sensors & Actuators B*, 106:614–618, 2005.

[546] H. Ji, D. Sander, A. Haas, and P.A. Abshire. A CMOS contact imager for locating individual cells. In *Int'l Symp. Circuits & Systems (ISCAS)*, pages 3357–3360, Kos, Greece, May 2006.

[547] T. Tokuda, A. Yamamoto, K. Kagawa, M. Nunoshita, and J. Ohta. A CMOS image sensor with optical and potential dual imaging function for on-chip bioscientific applications. *Sensors & Actuators A*, 125(2):273–280, February 2006.

[548] T. Tokuda, I. Kadowaki, .K Kagawa, M. Nunoshita, and J. Ohta. A new imaging scheme for on-chip DNA spots with optical / potential dual-image CMOS sensor in dry situation. *Jpn. J. Appl. Phys.*, 46(4B):2806–2810, April 2007.

[549] T. Tokuda, K. Tanaka, M. Matsuo, K. Kagawa, M. Nunoshita, and J. Ohta. Optical and electrochemical dual-image CMOS sensor for on-chip biomolecular sensing applications. *Sensors & Actuators A*, 135(2):315–322, April 2007.

[550] D. C. Ng, T. Tokuda, A. Yamamoto, M. Matsuo, M. Nunoshita, H. Tamura, Y. Ishikawa, S. Shiosaka, and J. Ohta. A CMOS Image Sensor for On-chip

in vitro and *in vivo* Imaging of the Mouse Hippocampus. *Jpn. J. Appl. Phys.*, 45(4B):3799–3806, April 2006.

[551] D. C. Ng, H. Tamura, T. Tokuda, A. Yamamoto, M. Matsuo, M. Nunoshita, Y. Ishikawa, S. Shiosaka, and J. Ohta. Real Time *In vivo* Imaging and Measurement of Serine Protease Activity in the Mouse Hippocampus Using a Dedicated CMOS Imaging Device. *J. Neuroscience Methods*, 156(1-2):23–30, September 2006.

[552] D. C. Ng, T. Tokuda, A. Yamamoto, M. Matsuo, M. Nunoshita, H. Tamura, Y. Ishikawa, S. Shiosaka, and J. Ohta. On-chip biofluorescence imaging inside a brain tissue phantom using a CMOS image sensor for *in vivo* brain imaging verification. *Sensors & Actuators B*, 119(1):262–274, November 2006.

[553] D.C. Ng, T. Nakagawa, T. Tokuda, M. Nunoshita, H. Tamura, Y. Ishikawa, S. Shiosaka, and J. Ohta. Development of a Fully Integrated Complementary Metal-Oxide Semiconductor Image Sensor-based Device for Real-time *In vivo* Fluorescence Imaging inside the Mouse Hippocampus. *Jpn. J. Appl. Phys.*, 46(4B):2811–2819, April 2007.

[554] D.E. Schwartz, E. Charbon, and K.L. Shepard. A Single-Photon Avalanche Diode Array for Fluorescence Lifetime Imaging Microscopy. *IEEE J. Solid-State Circuits*, 43(11):2546–2557, 2008.

[555] H.-J. Yoon, S. Itoh, and S. Kawahito. A CMOS Image Sensor With In-Pixel Two-Stage Charge Transfer for Fluorescence Lifetime Imaging. *IEEE Trans. Electron Dev.*, 561(2):214–221, February 2009.

[556] M.-W. Seo, K. Kagawa, K. Yasutomi, Y. Kawata, N. Teranishi, Z. Li, I.A. Halin, , and S. Kawahito. A 10 ps Time-Resolution CMOS Image Sensor With Two-Tap True-CDS Lock-In Pixels for Fluorescence Lifetime Imaging. *IEEE J. Solid-State Circuits*, 51(1):141–154, January 2016.

[557] S.H. Kim, S. Iwai, S. Araki, S. Sakakihara, R. Iino, and H. Noji. Largescale femtoliter droplet array for digital counting of single biomolecules. *Lab Chip*, 12(23):4986–4991, December 2012.

[558] K. Sasagawa, K. Ando, T. Kobayashi, T. Noda, T. Tokuda, S.-H. Kim, R. Iino, H. Noji, and J. Ohta. Complementary Metal-Oxide-Semiconductor Image Sensor with Microchamber Array for Fluorescent Bead Counting. *Jpn. J. Appl. Phys.*, 51(2S):02BL01, February 2012.

[559] Hironari Takehara, K. Miyazaki, T. Noda, K. Sasagawa, T. Tokuda, S.-H. Kim, R. Iino, H. Noji, and J. Ohta. A CMOS image sensor with stacked photodiodes for lensless observation system of digital enzyme-linked immunosorbent assay. *Jpn. J. Appl. Phys.*, 53:04EL02, 2014.

[560] K. Sasagawa, Hironari Takehara, M. Nagasaki, Hiroaki Takehara, T. Noda, T. Tokuda, H. Noji, and J. Ohta. Lensless CMOS Imaging Device for

Fluorescent and Non-Fluorescent Imaging Dedicated to Digital ELISA. *IEEJ Trans. Sensors and Micromachines*, 136(1):12–17, 2016. In Japanese.

[561] Hironari Takehara, M. Nagasaki, K. Sasagawa, Hiroaki Takehara, T. Noda, T. Tokuda, and J. Ohta. Micro-light-pipe array with an excitation attenuation filter for lensless digital enzyme-linked immunosorbent assay. *Jpn. J. Appl. Phys.*, 55:03DF03, 2016.

[562] T. Sakata, M. Kamahori, and Y. Miyahara. Immobilization of oligonucleotide probes on Si_3N_4 surface and its application to genetic field effect transistor. *Mater. Sci. Eng.*, C24:827–832, 2004.

[563] K. Hashimoto, K. Ito, and Y. Ishimori. Microfabricated disposable DNA sensor for detection of hepatitis B virus DNA. *Sensors & Actuators B*, 46:220–225, 1998.

[564] K. Dill, D.D. Montgomery, A.L. Ghindilis, and K.R. Schwarzkopf. Immunoassays and sequence-specific DNA detection on a microchip using enzyme amplified electrochemical detection. *J. Biochem. Biophys. Methods*, 2004.

[565] H. Miyahara, K. Yamashita, M. Takagi, H. Kondo, and S. Takenaka. Electrochemical array (ECA) as an integrated multi-electrode DNA sensor. *Trans.IEE Jpn.*, 121-E(4):187–191, 2001.

[566] A. Frey, M. Schienle, C. Paulus, Z. Jun, F. Hofmann, P. Schindler-Bauer, B. Holzapfl, M. Atzesberger, G. Beer, M. Frits, T. Haneder, H.-C. Hanke, and R. Thewes. A digital CMOS DNA chip. In *Int'l Symp. Circuits & Systems (ISCAS)*, pages 2915–2918, Kobe, Japan, May 2005.

[567] H. Nakazawa, H. Ishii, M. Ishida, and K. Sawada. A Fused pH and Fluorescence Sensor Using the Same Sensing Area. *Appl. Phys. Express*, 3:047001, 2010.

[568] H. Nakazawa, M. Ishida, and K. Sawada. Progressive-Type Fused pH and Optical Image Sensor. *Jpn. J. Appl. Phys.*, 49(4S):04DL04, 2010.

[569] H. Nakazawa, M. Ishida, and K. Sawada. Reduction of Interference Between pH and Optical Output Signals in a Multimodal Bio-Image Sensor. *IEEE Sensors Journal*, 11(11):2718 – 2722, November 2011.

[570] H. Nakazawa, M. Ishida, and K. Sawada. Multimodal bio-image sensor for real-time proton and fluorescence imaging. *Sensors & Actuators B*, 180:14–20, April 2011.

[571] A. W. Toga and J. C. Mazziota. *Brain Mapping: The Methods*. Academic Press, New York, NY, 2nd edition, 2002.

[572] A. Tagawa, A. Higuchi, T. Sugiyama, K. Sasagawa, T. Tokuda, H. Tamura, Y. Hatanaka, Y. Ishikawa, S. Shiosaka, and J. Ohta. Multimodal Complementary Metal-Oxide-Semiconductor Sensor Device for Imaging of

Fluorescence and Electrical Potential in Deep Brain of Mouse. *Jpn. J. Appl. Phys.*, 49:01AG02, January 2010.

[573] J. Ohta, Y. Ohta Hiroaki Takehara, T. Noda, K. Sasagawa, T. Tokuda, M. Haruta, T. Kobayashi, Y.M. Akay, and M. Akay. Implantable Microimaging Device for Observing Brain Activities of Rodents. *Proc. IEEE*, 105(1):158–166, January 2017.

[574] T. Tokuda, M. Takahashi, K. Uejima, K. Masuda, T. Kawamura, Y. Ohta, M. Motoyama, T. Noda, K. Sasagawa, T. Okitsu, S. Takeuchi, and J. Ohta. CMOS image sensor-based implantable glucose sensor using glucose-responsive fluorescent hydrogel. *Biomed. Opt. Express*, 5(11):3859–3870, 2014.

[575] T. Kawamura, K. Masuda, T. Hirai, Y. Ohta, M. Motoyama, Hironari Takehara, T. Noda, K. Sasagawa, T. Tokuda, T. Okitsu, S. Takeuchi, and J. Ohta. CMOS-based implantable glucose monitoring device with improved performance and reduced invasiveness. *Electron. Lett.*, 51(10):738–740, 2015.

[576] T. Tokuda, T. Kawamura, K. Masuda, T. Hirai, Hironari Takehara, Y. Ohta, M. Motoyama, Hiroaki Takehara, T. Noda, K. Sasagawa, T. Okitsu, S. Takeuchi, and J. Ohta. In-vitro long-term performance evaluation and improvement in the response time of CMOS-based implantable glucose sensors. *IEEE Design & Test*, 33:37–48, 2016.

[577] M. Haruta, C. Kitsumoto, Y. Sunaga, Hironari Takehara, T. Noda, K. Sasagawa, T. Tokuda, and J. Ohta. An implantable CMOS device for blood-flow imaging during experiments on freely moving rats. *Jpn. J. Appl. Phys.*, 53:04EL05, April 2014.

[578] M. Haruta, Y. Sunaga, T. Yamaguchi, Hironari Takehara, Y. Ohta, M. Motoyama, Hiroaki Takehara, T. Noda, K. Sasagawa, T. Tokuda, and J. Ohta. An implantable hemodynamic imaging device for observing the process of recovery from cerebrovascular disease. In *IEEE Annual Int'l Conf. Eng. Med. Bio. Soc. (EMBC)*, page FrFPoT2.36, Milan, Italy, 2015.

[579] M. Haruta, Y. Sunaga, T. Yamaguchi, Hironari Takehara, T. Noda, K. Sasagawa, T. Tokuda, and J. Ohta. Intrinsic signal imaging of brain function using a small implantable CMOS imaging device. *Jpn. J. Appl. Phys.*, 54:04DL10, April 2015.

[580] Hiroaki Takeahra, Y. Ohta, M. Motoyama, M. Haruta, M. Nagasaki, Hironari Takehara, T. Noda K. Sasagawa, T. Tokuda, and J. Ohta. Intravital fluorescence imaging of mouse brain using implantable semiconductor devices and epi-illumination of biological tissue. *Biomed. Opt. Express*, 6(5):1553, May 2015.

[581] H. Yawo, H. Kandori, and A. Koizumi, editors. *Optogenetics: Light-Sensing Proteins and Their Applications*. Springer, Tokyo, Japan, 2015.

[582] T. Tokuda, H. Kimura, T. Miyatani, Y. Maezawa, T. Kobayashi, T. Noda, K. Sasagawa, and J. Ohta. CMOS on-chip bio-imaging sensor with integrated micro light source array for optogenetics. *Electron. Lett.*, 48(6):312 – 314, March 2012.

[583] T. Tokuda, Hiroaki Takehara, T. Noda, K. Sasagawa, and J. Ohta. CMOS-Based Optoelectronic On-Chip Neural Inteface Devices. *IEICE Trans. Electron.*, E99-C(2):165 – 172, February 2016.

[584] M. Haruta, N. Kamiyama, S. Nakajima, M. Motoyama, M. Kawahara, Y. Ohta, A. Yamasaki, Hironari Takehara, T. Noda, K. Sasagawa, Y. Ishikawa, T. Tokuda, H. Hashimoto, and J. Ohta. Implantable optogenetic device with CMOS IC technology for simultaneous optical measurement and stimulation. *Jpn. J. Appl. Phys.*, 56:057001, March 2017.

[585] T. Kobayashi, M. Haruta, K. Sasagawa, M. Matumata, K. Eizumi, C. Kitsumoto, M. Motoyama, Y. Maezawa, Y. Ohta, T. Noda, K. Sasagawa, T. Tokuda, Y. Ishikawa, and J. Ohta. Optical communication with brain cells by means of an implanted duplex micro-device with optogenetics and Ca^{2+} fluoroimaging. *Sci. Rep.*, 6:21247, February 2016.

[586] G. Iddan, G. Meron, A. Glukhovsky, and P. Swain. Wireless capsule endoscopy. *Nature*, 405(6785):417, May 2000.

[587] R. Eliakim. Esophageal capsule endoscopy (ECE): four case reports which demonstrate the advantage of bi-directional viewing of the esophagus. In *Int'l Conf. Capsule Endoscopy*, pages 109–110, Florida, 2004.

[588] S.N. Hong, S.-H. Kang, H.J. Jang, and M.B. Wallace. Recent Advance in Colon Capsule Endoscopy: What's New? *Clinical Endscope*, 51:334–343, 2018.

[589] S. Itoh and S. Kawahito. Frame Sequential Color Imaging Using CMOS Image Sensors. In *ITE Annual Convention*, pages 21–2, 2003. In Japanese.

[590] S. Itoh, S. Kawahito, and S. Terakawa. A 2.6mW 2fps QVGA CMOS One-chip Wireless Camera with Digital Image Transmission Function for Capsule Endoscopes. In *Int'l Symp. Circuits & Systems (ISCAS)*, pages 3353–3356, Kos, Greece, May 2006.

[591] X. Xie, G. Li, X. Chen, X. Li, and Z. Wang. A Low-Power Digital IC Design Inside the Wireless Endoscopic Capsule. *IEEE J. Solid-State Circuits*, 40(11):2390–2400, 2006.

[592] S. Kawahito, M. Yoshida, M. Sasaki, K, Umehara, D. Miyazaki, Y. Tadokoro, K. Murata, S. Doushou, and A. Matsuzawa. A CMOS Image Sensor with Analog Two-Dimensional DCT-Based Compression Circuits for one-chip Cameras. *IEEE J. Solid-State Circuits*, 32(12):2030–2041, December 1997.

[593] K. Aizawa, Y. Egi, T. Hamamoto, M. Hatoria, M. Abe, H. Maruyama, and H. Otake. Computational image sensor for on sensor compression. *IEEE Trans. Electron Dev.*, 44(10):1724 – 1730, October 1997.

[594] Z. Lin, M.W. Hoffman, W.D. Leon-Salas, N. Schemm, and S. Balkr. A CMOS Image Sensor for Focal Plane Decomposition. In *Int'l Symp. Circuits & Systems (ISCAS)*, pages 5322–5325, Kobe, Japan, May 2005.

[595] A. Bandyopadhyay, J. Lee, R. Robucci, and P. Hasler. A 8 μW/frame 104 $\times$ 128 CMOS imager front end for JPEG Compression. In *Int'l Symp. Circuits & Systems (ISCAS)*, pages 5318–5312, Kobe, Japan, May 2005.

[596] T.B. Tang, E. A. Johannesen, L. Wang, A. Astaras, M. Ahmadian, A. F. Murray, J.M Cooper, S.P. Beaumont, B.W. Flynn, and D.R.S. Cumming. Toward a Miniature Wireless Integrated Multisensor Microsystem for Industrial and Biomedical Applications. *IEEE Sensors Journal*, 2:628–635, 2002.

[597] A. Astaras, M. Ahmadian, N. Aydin, L. Cui, E. Johannessen, T.-B. Tang., L. Wang, T. Arslan, S.P. Beaumont, B.W. Flynn, A.F. Murray, S.W. Reid, P. Yam, J.M. Cooper, and D.R.S. Cumming. A miniature integrated electronics sensor capsule for real-time monitoring of the gastrointestinal tract (IDEAS). In *Int'l Conf. Biomedical Eng. (ICBME) : "The Bio-Era: New Challenges, New Frontiers"*, pages 4–7, Singapore, December 2002.

[598] J.G. Linvill and J.C. Bliss. A direct translation reading aid for the blind. *Proc. IEEE*, 54(1):40–51, January 1966.

[599] J.S. Brugler, J.D. Meindl, J.D. Plummer, P.J. Salsbury, and W.T. Young. Integrated electronics for a reading aid for the blind. *IEEE J. Solid-State Circuits*, SC-4:304–312, December 1969.

[600] K. Motonomai, T. Watanabe, J. Deguchi, T. Fukushima, H. Tomita, E. Sugano, M. Sato, H. Kurino, M. Tamai, and M. Koyanagi. Evaluation of Electrical Stimulus Current Applied to Retina Cells for Retinal Prosthesis. *Jpn. J. Appl. Phys.*, 45(4B):3784–3788, April 2006.

[601] T. Watanabe, K. Komiya, T. Kobayashi, R. Kobayashi, T. Fukushima, H. Tomita, E. Sugano, M. Sato, H. Kurino, T. Tanaka, M. Tamai, and M. Koyanagi. Evaluation of Electrical Stimulus Current to Retina Cells for Retinal Prosthesis by Using Platinum-Black (Pt-b) Stimulus Electrode Array. In *Ext. Abst. Int'l Conf. Solid State Devices & Materials (SSDM)*, pages 890–891, Yokohama, Japan, 2006.

[602] A. Y. Chow, M. T. Pardue, V. Y. Chow, G. A. Peyman, C. Liang, J. I. Perlman, and N. S. Peachey. Implantation of silicon chip microphotodiode arrays into the cat subretinal space. *IEEE Trans. Neural Syst. Rehab. Eng.*, 9:86–95, 2001.

[603] A.Y. Chow, V.Y. Chow, K. Packo, J. Pollack, G. Peyman, and R. Schuchard. The artificial silicon retina microchip for the treatment of vision loss from retinitis pigmentosa. *Arch. Ophthalmol.*, 122(4):460–469, 2004.

[604] E. Zrenner. Will Retinal Implants Restore Vision? *Science*, 295:1022–1025, February 2002.

[605] D. Palanker, A. Vankov, P. Huie, and S. Baccus. Design of a high-resolution optoelectronic retinal prosthesis. *J. Neural Eng.*, 2:S105–S120, 2005.

[606] L. Wang, K. Mathieson, T.I. Kamins, J.D. Louding, L. Galambos, G. Goetz, A. Sher, Y. Mandel, P. Huie, D. Lavinsky, and D. Palanker. Photovoltaic retinal prosthesis: implant fabrication and performance. *J. Neural Eng.*, 9(4):046014–046025, 2012.

[607] C.-Y. Wu, P.-H. Kuo, P.-K. Lin, P.-C. Chen, W.-J. Sung, J. Ohta, T. Tokuda, and T. Noda. A CMOS 256-pixel Photovoltaics-powered Implantable Chip with Active Pixel Sensors and Iridium-oxide Electrodes for Subretinal Prostheses. *Sensors & Materials*, 30(2):193–211, 2018.

[608] A. Rothermel, L. Liu, N.P. Aryan, M. Fischer, J. Wuenschmann, S. Kibbel, and A. Harscher. A CMOS Chip With Active Pixel Array and Specific Test Features for Subretinal Implantation. *IEEE J. Solid-State Circuits*, 44(1):290–300, 2009.

[609] J. Ohta, T. Noda, K. Shodo, Y. Terasawa, Harut M, K. Sasagawa, and T. Tokuda. Stimulator Design of Retinal Prosthesis. *IEICE Trans. Electron.*, E100-C(6):523–528, June 2017.

[610] T. Noda, K. Sasagawa, T. Tokuda, Y. Terasawa, H. Tashiro, H. Kanda, T. Fujikado, and J. Ohta. Smart electrode array device with CMOS multi-chip architecture for neural interface. *Electron. Lett.*, 48(21):1328 – 1329, October 2012.

[611] T. Tokuda, Y. Takeuchi, Y. Sagawa, T. Noda, K. Sasagawa, Y. Terasawa, K. Nishida, T. Fujikado, and J. Ohta. Development and *in vivo* Demonstration of CMOS-Based Multichip Retinal Stimulator With Simultaneous Multisite Stimulation Capability. *IEEE Trans. Biomedical Circuits & Sytems*, 4(6):445–453, December 2010.

[612] T. Tokuda, K. Hiyama, S. Sawamura, K. Sasagawa, Y. Terasawa, K. Nishida, Y. Kitaguchi, T. Fujikado, Y. Tano, and J. Ohta. CMOS-Based Multichip Networked Flexible Retinal Stimulator Designed for Image-Based Retinal Prosthesis. *IEEE Trans. Electron Dev.*, 56(11):2577–2585, November 2009.

[613] A. Uehara, K. Kagawa, T. Tokuda, J. Ohta, and M. Nunoshita. A CMOS retinal prosthesis with on-chip electrode impedance measurement. *Electron. Lett.*, 40(10):582–583, March 2004.

[614] Y.-L. Pan, T. Tokuda, A. Uehara, K. Kagawa, J. Ohta, and M. Nunoshita. A Flexible and Extendible Neural Stimulation Device with Distributed

Multi-chip Architecture for Retinal Prosthesis. *Jpn. J. Appl. Phys.*, 44(4B):2099–2103, April 2005.

[615] A. Uehara, Y.-L. Pan, K. Kagawa, T. Tokuda, J. Ohta, and M. Nunoshita. Micro-sized photo detecting stimulator array for retinal prosthesis by distributed sensor network approach. *Sensors & Actuators A*, 120(1):78–87, May 2005.

[616] T. Tokuda, Y.-L. Pan, A. Uehara, K. Kagawa, M. Nunoshita, and J. Ohta. Flexible and extendible neural interface device based on cooperative multi-chip CMOS LSI architecture. *Sensors & Actuators A*, 122(1):88–98, July 2005.

[617] T. Tokuda, S. Sugitani, M. Taniyama, A. Uehara, Y. Terasawa, K. Kagawa, M. Nunoshita, Y. Tano, and J. Ohta. Fabrication and validation of a multi-chip neural stimulator for *in vivo experiments toward retinal prosthesis. Jpn. J. Appl. Phys.*, 46(4B):2792–2798, April 2007.

[618] A. Dollberg, H.G. Graf, B. Höfflinger, W. Nisch, J.D. Schulze Spuentrup, K. Schumacher, and E. Zrenner. A Fully Testable Retinal Implant. In *Proc. Int'l. Conf. Biomedical Eng.*, pages 255–260, Salzburg, June 2003.

[619] E. Zrenner. Subretinal chronic multi-electrode arrays implanted in blind patients. In *Abstract Book Shanghai Int'l Conf. Physiological Biophysics*, page 147, Shanghai, China, 2006.

[620] J. Ohta, T. Tokuda, Ke. Kagawa, Y. Terasawa, M. Ozawa, T. Fujikado, and Y. Tano. Large-scale Integration-Based Stimulus Electrodes for Retinal Prosthesis. In M.S. Humayun, J.D. Weiland, G. Chader, and E. Greenbaum, editors, *Artificial Sight*, chapter 8, pages 151–168. Springer, 2007.

[621] G.F. Poggio, F. Gonzalez, and F. Krause. Stereoscopic mechanisms in monkey visual cortex: binocular correlation and disparity selectivity. *J. Neurosci.*, 8(12):4531 – 4550, December 1988.

[622] D.A. Atchison and G. Smith. *Optics of the Human Eye*. Butterworth-Heinemann, Oxford, UK, 2000.

[623] B.A. Wandell. *Foundations of Vision*. Sinauer Associates, Inc., Sunderland, MA, 1995.

[624] B. Sakmann and O.D. Creutzfeldt. Scotopic and mesopic light adaptation in the cat's retina. *Pflögers Arch.*, 313(2):168–185, June 1969.

[625] J.M. Valeton and D. van Norren. Light adaptation of primate cones: An analysis based on extracellular data. *Vision Research*, 23(12):1539–1547, December 1983.

[626] V.C. Smith and J. Pokorny. Spectral sensitivity of the foveal cone photopigments between 400 and 500 nm. *Vision Research*, 15(2):161–171, February 1975.

[627] A. Roorda and D.R. Williams. The arrangement of the three cones classes in the living human eye. *Nature*, 397:520–522, February 1999.

[628] R.M. Swanson and J.D. Meindl. Ion-implanted complementary MOS transistors in low-voltage circuits. *IEEE J. Solid-State Circuits*, SC-7(2):146 – 153, April 1972.

[629] Y. Taur and T. Ning. *Fundamental of Modern VLSI Devices*. Cambridge University Press, Cambridge, UK, 1988.

[630] S.-C. Liu, J. Karmer, G. Indiveri, T. Delbrük, and R. Douglas. *Analog VLSI Circuits and Principles*. The MIT Press, Cambridge, MA, 2002.

[631] N. Egami. Optical image size. *J. Inst. Image Information & Television Eng.*, 56(10):1575–1576, November 2002.

[632] Scott Prahl. Optical Absorption of Hemoglobin. https://omlc.org/spectra/hemoglobin/.

[633] M.S. Patterson, B.C. Wilson, and D.R. Wyman. The Propagation of Optical Radiation in Tissue. II: Optical Properties of Tissues and Resulting Fluence Distributions. *Lasers in Meidal Science*, 6:379–390, 1991.

[634] G.M. Hale and M.R. Querry. Optical Constants of Water in the 200-nm to 200-μm Wavelength Region. *Appl. Opt.*, 12(3):555–563, March 1973.

Index

2K, 225
3D integration, 61, 79, 85
  – retinal prosthesis, 195
3D range finder, 9, 97, 130, 132, 137
4K, 225
8K, 225

α-Si, 55
aberration, 94
absorbance, 163
absorption, 55, 82, 84, 101
  – coefficient, 12, 24, 81, 87
  – hemoglobin, 183, 227
  – length, 13
  – oxidized hemoglobin, 227
  – spectrum in the tissue, 183
  – water, 183, 227
AC mode, 198
acceptor concentration, 14
accumulation mode, 2, 29, 130, 164
acetylcholine, 160
adaptive gain control, 112
ADAS: advanced driver assistance system, 107, 130, 137
ADC: analog-to-digital converter, 7, 48, 68, 107, 111, 113, 117, 175
  – 1-bit, 68, 143
  – bit serial, 129
  – column parallel, 48
  – cyclic, 48, 129
  – Delta-Sigma, 48
  – pixel-level, 48, 69, 76, 129
  – SAR: successive approximation register, 48
  – SS: single slope, 48, 50, 69, 76, 113
address event representation, 67, 70

Ag, 91
Ag/AgCl electrode, 174, 179
agarose gel, 179
Al, 91, 101, 178, 188, 194
  – pad, 194
  – wire, 178
Al_2O_3, 81
AM: amplitude modulation, 132
AMD: age-related macular degeneration, 193
AMI: amplified MOS imager, 4
antibody, 170
antigen, 170
antigen-antibody reaction, 170
APD: avalanche photodiode, 16, 18, 26, 75, 111, 119, 167
  – gain, 26
  – high-speed, 25, 26
  – multiplication, 27
  – SPAD, 27
  – VAPD, 28
APS: active pixel sensor, 4, 36, 38
  – 3T, 4, 6, 37, 38, 40, 57, 93, 143, 181
  – 4T, 4, 7, 36, 40, 42, 57, 144, 164
  – current mode, 62
  – differential, 110
  – modified 3T, 176
  – PG, 4, 133
  – speed, 53
AR: augmented reality, 155, 156
ASIC Vision, 7
ASP: angle sensitive pixel, 97
asynchronous operation, 68
Au, 178
auto focus, 92
auto-zero operation, 47

avalanche, 18
   – breakdown, 18, 21, 26
   – breakdown voltage, 27
   – PCD, 28

band bending, 24
bandgap, 15, 20, 94
   – Si, 15, 78
bandwidth, 23
BASIS: base-stored image sensor, 4
Bayer pattern, 55
BiCMOS technology, 24, 26
binocular method, 142
bio-compatibility, 194
bio-marker, 170, 171
biomorphic digital vision chip, 7
bioscience, 165
biotechnology, 165
bipolar cell, 193
birefringence, 98
black Si, 82
blood flow, 183, 184
blood vessel, 181, 183, 184
Bluetooth, 147, 157
body transconductance, 45
body-effect coefficient, 222, 223
BOX: buried oxide, 79, 80
BPD: buried photodiode, 34, 109
BPSK: binary phase shift keying, 192
brain imaging, 185
BSI: backside illumination, 8, 28, 60, 79, 84, 85, 94, 102, 194
built-in potential, 19, 30, 31, 80
bullseye structure, 91
bump, 82, 189
   – Pt/Au stacked, 202
burst mode, 118

camera-on-a-chip, 107
capacitance
   – FD, 41, 51
   – PD, 39, 40, 51
capacitive coupling measurement, 173
CAPD: current-assisted photonic demodulator, 67

capsule endoscope, 190, 191
CCD: charge coupled device, 1, 3, 42, 57, 119
CDS: correlated double sampling, 4, 11, 41, 45, 46, 52, 111, 113, 116
   – chip-level, 48
   – digital, 50, 113
channel length modulation, 222
charge amplifier, 2, 37
charge balance, 201
charge transfer
   – efficiency, 92
   – incomplete, 42
chemical imaging sensor, 163
chemical reaction, 162
chip-level processing, 107
chirality, 160
ChR2: Channel Rhodopsin 2, 188
CIF: common intermediate format, 48, 225
clamp capacitor, 47
CMD: charge modulated device, 4
CMS: correlated multi-sampling, 111
CNN: cellular neural network, 7
coded aperture camera, 103
colon, 191
color filter, 54, 86, 88, 90, 91
   – CMY, 55
   – inorganic film, 55, 56
   – interference flter, 56
   – near infrared, 56
   – on-chip, 29, 54
   – organic film, 55
   – photonic crystal, 56
   – RGB, 54
   – white pixel, 55
column-level processing, 107
column-parallel processing, 107, 111, 114, 117
comparator, 50, 69, 77
complete depletion, 4, 35, 40
completely depleted voltage, 35
compound eye, 95
   – apposition eye, 95
   – arthropod, 95

– artificial, 95
– crustacean, 95
– insect, 92, 95
– superposition eye, 95
– TOMBO, 96
conduction band, 20, 31, 219
constant current mode, 45
constant voltage mode, 45
conversion gain, 40, 41, 111
correlation method, 131
cortical stimulation, 193
counter electrode, 179
CR time constant, 24, 53
crosstalk
– diffusion carrier, 79
– optical, 84
CTIA: capacitive transimpedance amplifier, 37, 62, 65
Cu, 101
current copier, 62
current mirror, 26, 62, 197
– mirror ratio, 62
– mismatch, 62
current mode, 26, 61, 62
– -ADC, 62
– accumulation mode, 62
– APS, 62
– current copier, 62
– direct mode, 62
– memory, 62
curved image sensor, 94
CV: cyclic voltammetry, 178

DAC: digital-to-analog converter, 48, 49
dangling bond, 23
dark current, 4, 20, 21, 70, 80, 94, 108, 109, 116
– bias voltage dependence, 22, 109
– PCD, 28
– temperature dependence, 22
data rate, 114
DDS: double delta sampling, 48
decoder, 11, 43
deep level, 34

deep n-well, 75
demodulation, 9, 66, 130, 139, 153
demosaicking, 55
depletion
– region, 20, 89
– width, 18, 19, 21, 24, 25
– voltage dependence, 30
dichroic mirror, 56
dielectric constant, 19
differential amplifier, 47, 197
diffractive optics, 97, 142
diffusion, 14, 66, 119, 125
– carrier, 20, 53, 79
– coefficient, 17, 24
– current, 17, 20, 223
– length, 14, 17, 34, 53, 151
– time, 24
digital ELISA, 170
digital mode, 61
DNA
– hybridization, 175, 178
– identification, 173
– probe, 176
donor concentration, 14
DPS: digital pixel sensor, 69, 76
DPSS: divisional power supply scheme, 197
DR: dynamic range, 40, 52
– definition, 52
– optical, 52
– output, 52
drain current, 221
DRAM: dynamic random access memory, 7, 85, 129
drift, 66, 119, 155, 168
DSC: digital still camera, 1
DSLR: digital single lens reflex, 1
dual PD, 126
dual sampling, 127
dual sensitivity, 126

Early effect, 222
– Early voltage, 222
EEG: electroencephalography, 181
effective density of states, 20

Einstein relation, 120
EIS: electrolytc-insulator-semiconductor,
        163, 164
electrical stimulation, 188, 195
electro-physiological   measurement,
        202
electrochemical measurement, 176
electrode
    – Ag/AgCl, 179
    – Al, 178, 188
    – Au, 178
    – Pt, 188, 194
    – recording, 188
    – stimulation, 194, 197
electrolysis, 198
electrolyte, 188
electron density, 31
electron-hole pair, 12, 28
electrophysiological measurement, 189
ELISA: enzyme-linked
        immuno-sorbent   assay,
        170
EMCCD: electron multiplying CCD,
        108
EMI: electro-magnetic   interference,
        147
endoscope, 190
enzyme, 170
ep-retina stimulation, 193
epithelium, 198
epoxy, 178
esophagus, 191
excess carrier density, 23
excitation light, 171, 185
extinction ratio, 99
eye
    – animal, 94
    – human, 122
eye-safe wavelength region, 83

F number, 94
FA: factory automation, 130, 137
FD: floating diffusion, 4, 40, 42, 47,
        111, 116, 126, 127, 133, 164,
        165, 167, 179

Fermi energy, 219
Fermi level, 31, 35
FF: fill factor, 28, 36, 42, 84, 108, 114,
        125, 130
FGA: floating gate array, 4
filter
    – interference, 105
    – absorption, 105
    – interference, 105
    – surface plasmon, 105
    –emission, 105
FlatCam, 103
flexible substrate, 185, 198
FLIM: fluorescence lifetime imaging
        microscopy, 28, 167, 169
flip-chip bonding, 82
flow cytometer, 104
FlowSign Light, 155, 156
fluorescence, 181, 185
    – contact imaging, 105
    – decay time, 167, 169
    – filterless detection, 89
    – lensless imaging, 105
    – lifetime, 169
    – onchip imaging, 169
    – optical microscope, 165, 171
fMRI: functional magnetic resonance
        imaging, 181
focal plane image processing, 1
FOV: field-of-view, 94, 95, 104, 105,
        142
fovea, 92
FPA: focal plane array, 80
FPN: fixed pattern noise, 4, 22, 50, 58,
        62, 64
    – column, 36, 48, 50
    – pixel, 50
    – PTr, 26
frame difference, 158
frame memory, 119, 129, 130
frame rate
    – video, 157
Frankel-Poole current, 21
free-space optical communication, 145,
        146

Fresnel zone, 103
FSI: front-side illumination, 84
FTTH: fiber-to-the-home, 24
full well capacity, 40, 52, 116, 125, 135

G–R current: generation–recombination current, 21
g–r current: generation–recombination current, 21
GaAs, 29
gain
  – VAPD, 28
  – APD, 26
  – PCD, 28
  – PTr, 26
gamma, 77
GaN, 29
ganglion cell, 193, 202
gate oxide, 58
Ge, 61, 78, 82, 83
Geiger mode, 26, 27, 75, 139
gesture recognition, 137
GFP: green fluorescence protein, 105
global shutter, 59, 114, 115
  – memory node, 116
glucose sensing, 183
GS: global shutter–6T-type, 116
guard ring, 75
GW: gate window, 168

HBT: hetero-structure bipolar transistor, 83
HD: high definition, 225
hemodynamic signal, 184, 189
hemoglobin, 184, 227
  – conformation change, 184
HgCdTe, 82
high frame rate, 114
high-definition television: HDTV, 117
high-speed, 76, 114
  – imaging, 167
  – readout, 158
hippocampus, 186
histogram equalization, 77

HOVI: hyper omni vision, 92
human interface, 145
human visual processing system, 2, 5, 7
hyperbolic mirror, 92

I–V converter, 4, 48
ICT: information and communication technology, 145
ID cam, 155
ID map table, 158
IEEE 802.15.7, 147
illuminance, 68, 209
image blurring, 14
image lag, 42, 125
image tube, 2
impact ionization, 21
implantable medical devices, 79
implantation, 181
impurity concentration, 53
in vitro, 202
in vivo, 185
in vivo imaging, 182
in vivo window, 183, 227
in-pixel amplification, 4
infrared, 28
infrared image sensor, 80
  – cooled type, 82
  – uncooled type, 80
InGaAs, 82, 83
InSb, 82
interface state, 23, 34
interference film, 56
interference filter, 171
intraocular implantation, 193
intrinsic carrier concentration, 19, 21
intrinsic optical signal, 181, 184, 227
  – imaging, 183
inverted pyramidal structure, 81
inverter, 70, 75, 111
ionization coefficient, 21
IoT: Internet of Things, 156
IrDA, 147, 157
ISFET: ion sensitive field effect transistor, 163, 164
isolation layer, 58

ITRS: International Technology Roadmap, 7
ITS: intelligent transportation system, 107

Kirchhoff's current law, 62
kTC noise, 4, 23, 36, 39, 41, 46, 47, 51

LAN: local-area network, 147
LAPS: light-addressable potentiometric sensor, 163
large intestine, 191
latch, 43
lateral overflow integration capacitor, 126
LCP: liquid crystal polymer, 98
LED: light-emitting diode, 185, 188
 – blue, 145, 147
 – white, 145, 147, 191
 – blue, 189
 – display, 156
 – green, 183, 185, 189
 – multi-color, 146, 155
 – red, 185
LEFM: lateral electric field charge modulator, 67
lensless imaging, 103
LIDAR: light detection and ranging, 137, 139
lifetime, 14
 – deep level, 21
 – PCD, 28
light adaptation, 201
light field camera, 96
light section method, 142, 143
LinkRay, 155, 156
LOC: lab-on-chip, 192
lock-in pixel, 66, 131, 133, 139, 168
LOCOS: local oxidation of silicon, 58
log sensor, 57, 64, 125, 144, 197
 – -accumulation mode, 64
logarithmic response, 127
low light imaging, 108, 109
low-pass filter, 201
low-power operation, 68

low-voltage operation, 68

MAPP, 7
massively parallel-processing, 5
MEG: magnetoencephalography, 181
MEMS: micro-electro-mechanical systems, 80, 192
metal wire grid, 90, 98, 163
 – Al, 91
 – W, 91
micro-bump, 85
micro-chamber, 171, 172
micro-fluidic device, 172
micro-lens, 60, 84, 96, 97
 – digital, 142
 – split, 92
μTAS: micro total analysis system, 192
mid-gap energy, 219
MIMD: multiple instruction and multiple data, 7
minority carrier, 13
mobile application, 79
mobility, 14
modulated light, 130
Moore's law, 7
MOS capacitor, 25, 164, 219
 – accumulation mode, 219
 – depletion mode, 219
 – inversion mode, 220
MOSFET
 – bulk, 79
 – cutoff region, 221
 – depletion, 221
 – enhancement, 221
 – linear region, 221
 – liner region, 221
 – saturation region, 221, 222
 – strained, 83
 – sub-threshold region, 221, 222
   – linear region, 223
   – saturation region, 224
motion capture, 130
MSM: metal-semiconductor-metal
 – photo-detector, 29
 – GaAs, 29, 67

– GaN, 29
multi-modal
– functions, 173
– measurement, 163
– sensing, 192
multi-path structure, 81
multi-resolution, 45
multiple junctions, 88
multiple micro-chip based stimulator, 198
multiple sampling, 127, 128
multiplier, 131
MWIR: mid-wavelength infrared, 82

nano-deflector, 91
nano-splitter, 91
nano-structure, 82
– deflector, 91
– splitter, 91
near-sensor image processing, 7
nearest neighboring operation, 77
Nernst equation, 164
neural cell, 189
NIR: near infrared, 13, 53, 56, 81–83, 98, 196, 198, 203
– cut filter, 14
noise, 175
- input-referred, 112
– 1/f, 52, 109, 110
– background, 130
– cancellation, 130
– common mode, 110
– flicker, 109
– floor, 52
– Johnson, 23
– kTC, 4, 23, 36, 39–41, 46, 47, 51
– log sensor, 125
– Nyquist, 23
– random, 4
– reset, 130
– RTS, 52
– shot, 22, 52, 73
– sub-electron, 111
– thermal, 22, 23, 51
non-recombination center, 23

NSIP: near-sensor image processing, 7

OCC: optical camera communication, 146
OFD: overflow drain, 116, 124, 125, 127, 155
– lateral, 126
ommatidium, 95, 96
on-chip color filter, 185, 191
on-chip compression, 192
on-chip detection, 166
ON-resistance, 51
OPA: operational amplifier, 47
open circuit voltage, 17
opsin, 188
Optacon, 192
optical absorbance, 163
optical activity, 160, 163
optical beacon, 156
optical crosstalk, 55, 59
– CCD, 59
– CMOS image sensor, 59
optical fiber, 146
optical fiber communication, 24, 26
optical format, 225
optical ID tag, 147, 155, 157, 158
optical pilot signal, 158
optical response angle, 84
optical rotation, 160
– angle, 160, 161
optical stimulation, 9, 188
optical wireless communication, 146
– indoor, 147, 148
– near distance, 147
– outdoor, 147
optical wireless LAN, 24
OptNavi, 155, 156
optogenetics, 188
organic photo-conductive film, 87
OT: optical topography, 181

parallax, 143
PASIC, 7
patch-clamp, 178
pattern recognition, 2

PbS, 82
PCD: photo-conductive detector, 2, 16, 28
  – avalanche, 28
  – dark current, 28
  – gain, 28
  – gain-bandwidth, 28
  – lifetime, 28
  – organic film, 87
PD: photodiode, 16
  – buried, 34, 109
  – Ge, 83
  – near zero bias, 109
  – pinned, 4, 34, 41, 42, 109, 110
  – solar cell mode
    – retinal prosthesis, 195
  – speed, 24
  – tandem, 196
PDAF: phase-difference detection auto focus , 92
PDMS: polydimethylsiloxane, 172
PET: positron emission tomography, 181
PFM: pulse frequency modulation, 67, 69, 80, 81, 109, 130
  – retinal prosthesis, 198
PG: photo-gate, 4, 16, 25, 89, 90, 126, 179
pH, 9, 160, 164
  – imaging, 163, 179
photo-conductivity, 28
photo-current, 15, 17, 26
photo-detection, 12
photo-generated carrier, 12, 28
photo-receiver, 24
photo-receptor, 92, 193, 200
photo-sensitivity, 15, 80, 122
  – spectrum, 88
photonic crystal, 56, 90, 98
photopic, 215
photoscanner, 2
Picalico, 155
Picapicamera, 156
PicoCam, 103
pinhole camera, 103

pinning voltage, 35, 120
pixel pitch, 5
pixel rate, 114
pixel sharing, 4, 42, 59, 116
  – FD driving technique, 42
pixel-level processing, 7, 85, 108
pixel-parallel processing, 85, 108
plenoptic camera, 96
PM: phase modulation, 132
PMD: photo mixer device, 67
pn-junction
  – capacitance, 24
  – junction shape, 30
  – linear junction, 30
  – step junction, 30
pn-junction diode, 17
  – forward current, 17
  – ideal factor, 17
  – thermally isolation, 80
pn-junction photodiode
  – junction capacitance, 30
POC: point-of-care, 105
polarimeter, 161
polarimetric imaging, 98
polarization, 9, 90, 98
  – detection, 163
  – rotation angle, 163
polarizer, 161
poly-silicon, 4, 26, 58, 219
  – polycide, 58
polyimide, 185
post-processing, 96
potential description, 31
potential imaging, 9, 173, 175
potential profile control, 88, 185
potential well, 31
PPD: pinned photodiode, 4, 34, 41, 42, 109, 116
PPS: passive pixel sensor, 2, 36, 57, 127
processing
  – chip-level, 5
  – column level, 5
  – column parallel, 7
  – pixel-level, 5
programmable artificial retina, 6

projection, 69
proton, 163, 164
PSL: parasitic light sensitivity, 116
Pt, 188, 194
PTr: photo-transistor, 2, 4, 5, 16, 26, 62
– base current gain, β, 26
– base width, 26
– FPN, 26
– lateral, 80, 81
– parasitic transistor, 26
PtSi, 82
pulse mode, 61
pulse modulation, 67
PVA: polyvinyl alcohol, 98
PWM: pulse width modulation, 7, 67, 68, 77, 130, 143

QCIF: quarter common intermediate format, 173, 225
quantum dot, 82
quantum efficiency, 15, 18, 20, 24, 90
QVGA: quarter video graphics array, 48, 144, 192, 225
QWIP: quantum-well infrared photodetector, 82

ramp waveform, 69
random access, 44
range finder, 137
readout circuit, 4, 11, 28, 45
reset, 40, 51
– hard reset, 40
– soft reset, 40
reset transistor, 39, 41, 42, 44, 46, 51, 70, 109, 133
resistive network, 5, 6
resolution, 225
resonant oscillation, 91
retina, 193, 213
retinal cell, 196–198, 203
retinal prosthesis, 192
– epi-retinal space, 194
– sub-retinal space, 195
retinotopy, 193
RF, 146, 191, 192

RGB, 86
RGL: reset gate less, 111
robot vision, 137
robotics, 130
ROC: readout circuitry, 29
ROI: region-of-interest, 115, 155, 157, 158
ROIC: readout integrated circuits, 82
rolling shutter, 56, 59, 114, 156, 191
RP: retinitis pigmentosa, 193
RSG: ramp signal generator, 50
RTS: random telegraph signal, 52

S/H: sample and hold, 11, 45, 47, 128, 151
– capacitance, 45
saline solution, 174, 179
sapphire, 78, 81
saturation current, 17
saturation detection, 130
saturation velocity, 24
scanistor, 2
scanner, 2, 11, 43, 94
Schmitt trigger, 70
Schottky barrier photodetector, 82
scotopic, 215
SDM: spatial division multiplexing, 150
select transistor, 37, 38, 41, 45, 47
self-reset, 73
self-reset imaging, 187
sensing electrode, 173
sensitivity
– spectrum curve, 19
sensor network, 79
shadow imaging, 104
shift register, 11, 43
Si_3N_4, 20, 91, 173, 194
SiGe, 61, 78, 82, 83
silicide, 79
silicon retina, 5
silicone, 174
SIMD: single instruction and multiple data, 7
SiO_2, 20, 56, 78–80, 84, 219
SIT: static induction transistor, 4

SLM: spatial light modulator, 104
small intestine, 191
smear, 36
SNR: signal-to-noise ratio, 4, 23, 36, 39, 42, 52, 55, 67, 69, 74, 107, 111, 112, 116, 119, 127, 130, 165, 200
– dip, 128, 129
SoC: system-on-chip, 7, 190, 192
SOI: silicon-on-insulator, 61, 78, 79, 85
solar cell, 196
solar cell mode, 17
somato-sensory cortex, 184
SOS: silicon-on-sapphire, 61, 78, 81
– retinal prosthesis, 194
source follower, 4, 38, 45, 61, 65, 110, 119, 164, 173
– transistor, 45
– voltage gain, 45
source follower transistor, 87
SPAD: single photon avalanche diode, 27, 28, 75, 111, 119, 139, 168
– photon counting mode, 75
– photon timing mode, 75
– quench, 27
spectral matching, 132
SRH: Shockley–Read–Hall recombination, 21
stacked capacitor, 126
stacked CMOS image sensor, 57, 78, 85, 117, 195
stacked PDs, 171
stereo-vision method, 142
STI: shallow trench isolation, 28, 58, 79
stimulus current, 195
– biphasic, 201, 203
Stokes shift, 105
stomach, 191
stray capacitance, 53
strong inversion, 220
structured light method, 142, 143
STS: suprachoroid   transretinal stimulation, 193, 198
sub-retina stimulation, 193
sub-threshold

– current, 109, 222
– region, 5, 64
– slope, 223
sub-threshold operation, 125
sub-wavelength structure, 90
– aperture, 90
super HARP, 29, 108
surface leak current, 21
surface plasmon, 90, 91
surface potential, 164, 220
surface recombination, 20, 23
– rate, 21, 23
surface state, 23, 34
surveillance, 92
SVGA: super video graphics array, 225
SWIR: short-wavelength infrared, 82
switched capacitor, 201

target tracking, 45, 130
TCSP: time-correlated single photon counting, 168
TDC: time-to-digital counter, 75
tensile stress, 94
TFT: thin film transistor, 2, 89
thermal generation, 21
threshold voltage, 164
TIA: transimpedance amplifier, 150
time-to-saturation, 67
TiO_2, 56, 98
TOF: time-of-flight, 25, 66, 114, 118, 130, 139
– direct, 139
– indirect, 139
TOMBO, 96
transconductance, 45
transfer gate, 4, 42, 47, 133
transistor isolation, 58
transit time, 24
– depletion region, 24
triangulation, 142
TSL: transversal signal line, 36
TSV: through silicon via, 85
tungsten light shield, 116
tunnel current, 21, 109
– band-to-band, 21

– trap-assisted, 21
two accumulation method, 133

UHS: ultra-high-speed, 114, 118
ultra-fast photo-detector, 29
ultra-low-light detection, 26
unity-gain buffer, 178
UV: ultra violet, 28, 29
UXGA: ultra extended graphics array, 225

valence band, 20, 219
VAPD: vertical avalanche photodiode, 28, 111, 168
– gain, 28
variable photosensitivity, 201
VGA: video graphics array, 45, 144, 225
virus, 170, 171
visible light communication, 155
vision chip, 1, 5, 76
– analog, 7
– digital, 7
visual remote controller, 156
VLC: visible light communication, 145, 147
VLSI Vision, 7
voltage follower, 178
voltage-controlled   current measurement, 178

W, 91
– light shield, 116
WDM: wavelength   division multiplexing, 146, 152, 155
weak inversion, 220
well-capacity, 41
– adjustment, 125
wide angle camera, 96
wide dynamic range, 7, 64, 65, 68, 70, 76, 107, 122, 144, 197
work electrode, 179

X-ray, 28
XGA: extended graphics array, 225